NATIONAL GEOGRAPHIC

Exploring Science

Program Consultants

Randy L. Bell, Ph.D.

Malcolm B. Butler, Ph.D.

Kathy Cabe Trundle, Ph.D.

Judith S. Lederman, Ph.D.

Center for the Advancement of Science in Space, Inc.

Welcome to
Exploring Science

Nature of Science

Physical Science 20

Forces and Interactions

Life Science

Interdependent Relationships in Ecosystems

Life Science (continued)

Earth Science

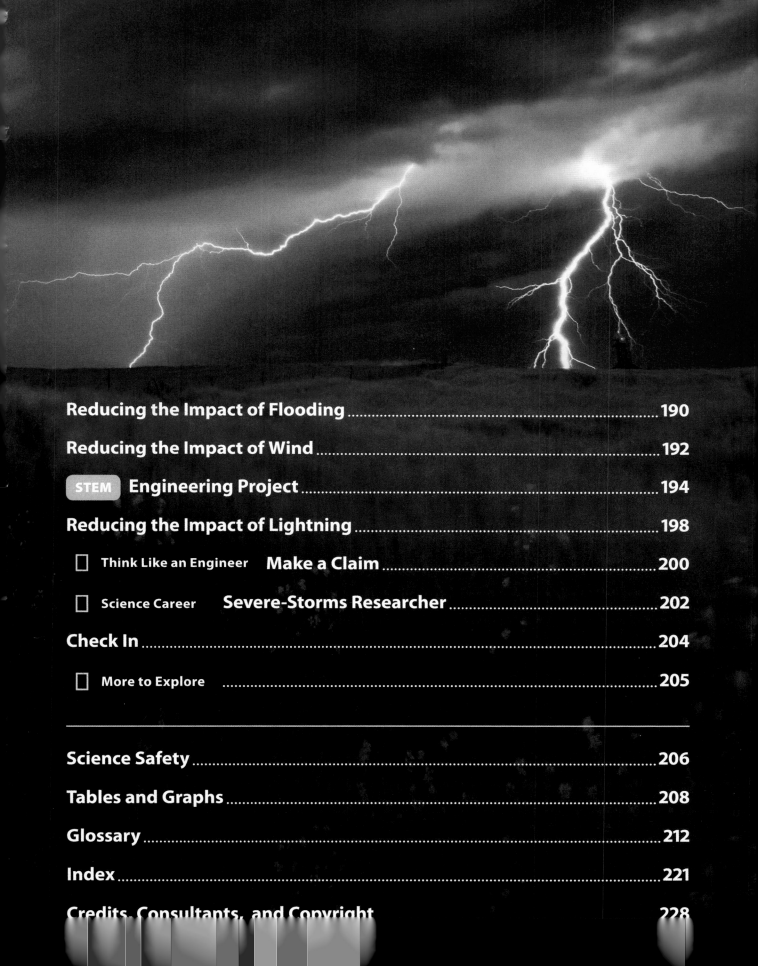

Andrés Ruzo is a geoscientist and a National Geographic Explorer. Andrés studies Earth's heat. Deep inside Earth it is very hot. Some of that heat moves toward Earth's surface. Andrés is curious. Could people use that heat for energy?

The shaman and local people have shared ancient knowledge of the Boiling River with Andrés. This plays a key role in his research.

Welcome to Exploring Science!

Hello, explorers! I am Andrés Ruzo. Welcome to *Exploring Science*! Together let's investigate how scientists answer questions and make new discoveries.

My research focuses on Earth's heat. This includes geothermal features like volcanoes and hot springs. When I was a boy, I grew up in Peru. My grandfather told stories of the sacred Boiling River of the Amazon. He said the water is hot enough to boil animals alive. Remembering that story as an adult and a scientist made me ask some questions. How could a river like this, far from any volcanoes, exist? I headed to the jungle to find out. I found the river and the local people who live near it and use its water every day. I met a local shaman, a healer and protector of the river. He gave me his blessing and I became the first geoscientist ever allowed to study the Boiling River. I got to work answering my questions.

Keeping a Science Notebook

📓 My Science Notebook

One tool I use in my research is a science notebook. I keep records of observations, measurements, and other data. I look for patterns and study the evidence. From these, I make predictions, explanations, and conclusions. In this program, you will learn how scientists and engineers ask questions and solve problems. And you can keep your own science notebook. Here are some ways to use your notebook. You or your teacher may have more ideas.

- Define and draw science words and main ideas.
- Label drawings. Include captions and notes to explain ideas.
- Collect objects, such as photos and magazine or newspaper clippings.
- Add tables, charts, or graphs to record observations and data.
- Record evidence for explanations and conclusions.
- Think about what you've done and learned. Ask new questions.

Look at the notebook examples for some ideas. Now it's time to set up your own science notebook!

A science notebook is one of a scientist's most useful tools.

Andrés's notebook includes calculations and graphs.

▼ Label drawings. Use captions to explain main ideas.

▼ Add tables, graphs, and maps to your science notebook.

Electromagnet

a battery labeled with positive and negative terminals

wire at negative terminal

wire at positive terminal

lug nut

wire wound around bolt

Electricity flows from the battery through the wire around the nail and back to the battery. The bolt becomes a magnet and picks up the lug nut.

Weather Prediction

A warm front brings warm weather. There is a warm front in the northwest. A cold front brings cooler weather. There is a cold front across the center of the country. Weather maps are important tools for predicting weather.

▶ You can use your notebook to think about what you've done and learned.

1. I thought like a scientist when I did research to obtain information on ways that living things change during their lives. I made a model of a life cycle.

2. I thought like an engineer when I designed a new use for magnets. Then I built and tested my design.

3. I learned the difference between weather and climate. I learned that climate is average weather over time.

Set up Your Science Notebook

Use your science notebook every time you study science. Here are a few suggestions to make your notebook unique and easy to use. Your teacher may have more instructions. Use your own ideas, too!

- Design a cover. Include something you like about science or something you would like to learn.

- In the front of your notebook, write "Table of Contents." Leave some blank pages. For each entry, write the date, title, and a page number.

- Organize your notebook. Add tables and graphs. Label everything carefully. Write the date on every new entry.

- Keep your science notebook in a safe place. You'll want it to last, so you can see how much you have learned.

▼ Design a cover that is all about science and you! Make a table of contents on the second page. Add information as you read and investigate.

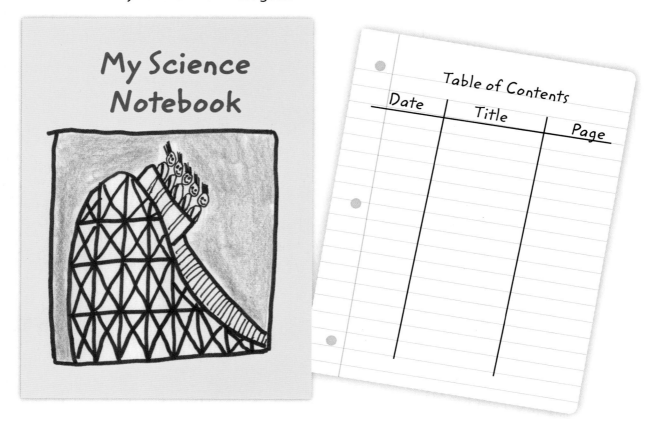

Why Explore Science?

Science is a way of learning about nature. It is a way of discovering and explaining how things work the way they do. And when you learn more about one area of science, you can often apply it to another. The basic laws of nature are the same everywhere in the universe.

You can use the same practices that scientists use to learn about nature. When you question, observe, explore, and explain the natural world, you are acting like a scientist. You can learn how science is used to solve important problems.

Look at the pages ahead. In *Exploring Science,* you will learn more about what science is and what it isn't. You will read about different ways that scientists work and one thing that all scientists rely on—evidence. As you look at the scientists, fossils, and study sites on the next pages, what questions do you have? Write them in your notebook!

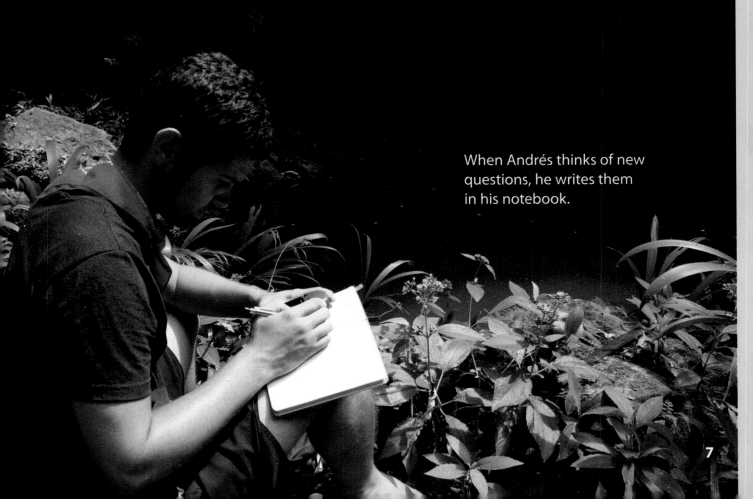

When Andrés thinks of new questions, he writes them in his notebook.

Nature of Science

Not all science happens in a lab. National Geographic Explorer Paul Sereno and his teammate are paleontologists. Paul led a team on an expedition to Niger. They made many discoveries about Earth's past. Here, Paul points to the largest ancient human graveyard ever found in the Sahara. Science can be exciting and full of discovery. But what is science, really?

What Is Science?

Science is a way of knowing about the natural world. It is also the knowledge itself. It is a body of evidence that builds over time. **Evidence** forms as scientists make sense of new information. Inference is also an important part of creating knowledge. Scientists' conclusions are **inferences** based on new observations and what is already known at the time.

People have not always had all the tools we have today. They did not have the collection of evidence we have today, so their inferences were different. When they tried to make sense of the world, they made inferences based on the facts they had at the time.

For example, the ancient Greeks may have found fossils of large animals that had a hole in the center of the skull. No living animals had such skulls. To make sense of these strange objects, they may have constructed myths about a Cyclops, a monster with one eye in the center of its head. We now know these fossils are the remains of extinct elephants. The hole is where its trunk was. As new discoveries are made and shared, science grows. Better inferences are made. And we can understand nature more clearly and deeply.

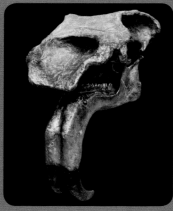

The mythical Cyclops has an eye in the middle of its head. Its story may have come from elephant fossils like this one.

DCI LS4.A: Evidence of Common Ancestry and Diversity. Some kinds of plants and animals that once lived on Earth are no longer found anywhere. (3-LS4-1)

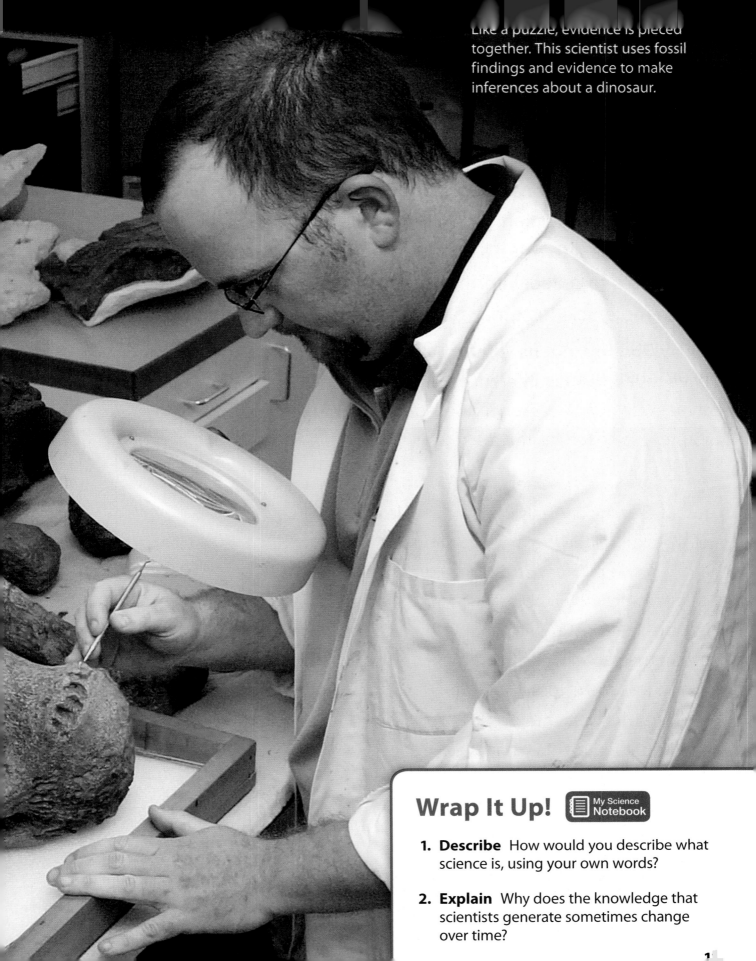

Like a puzzle, evidence is pieced together. This scientist uses fossil findings and evidence to make inferences about a dinosaur.

Wrap It Up! 📓 My Science Notebook

1. **Describe** How would you describe what science is, using your own words?

2. **Explain** Why does the knowledge that scientists generate sometimes change over time?

1

How Do Scientists Work?

There are many ways scientists can build evidence. One way is to conduct investigations. An **investigation** is a process in which scientists ask a question, plan a procedure to make observations, and gather data to make a conclusion. To **observe** means to use your senses to collect information about the world around you. For example, scientists may plan a fossil dig. But they don't just start digging. They carefully plan an investigation to make and record their observations, which can be analyzed for patterns.

In science, a model explains or predicts nature. These dinosaur models can help explain how animals lived long ago.

DCI LS4.A: Evidence of Common Ancestry and Diversity. Fossils provide evidence about the types of organisms that lived long ago and also about the nature of their environments. (3-LS4-1)
NS Scientific Knowledge Is Based on Empirical Evidence. Science findings are based on recognizing patterns. (3-LS1-1), (3-PS2-2)
NS Scientific Investigations Use a Variety of Methods. Science investigations use a variety of methods, tools, and

An experiment is a special type of investigation. An **experiment** is a fair test, a process in which scientists control variables to find out how variables are related to each other. A **variable** is a factor in an experiment that may change. By controlling variables, scientists can find out if one variable causes the other.

Scientists also use models. In science, a **model** is an inference. Models can be used to explain or predict natural systems. For example, when scientists find a fossil of an animal that has not been described before, they can make a model of the organism based on their observations. The model can help explain how the bones fit together, how the animal moved, or how it lived. The model can be used to make inferences about past environments.

Wrap It Up! My Science Notebook

1. **Explain** What is the difference between an investigation and an experiment?

2. **Describe** How do scientists use models to generate knowledge?

Who Are Scientists?

People from all different cultures and backgrounds study science. Most scientists work in teams. Some teams study living things or the environments in which they live. Other scientists study processes that occur on Earth, such as weather or erosion. Still other scientists study energy, forces, and materials that are used to make the technologies we use every day.

No matter what a scientist's background or what he or she studies, science has certain characteristics. Curiosity and creativity are valuable in science. So is attention to detail. Scientists ask questions that can be answered by conducting investigations, making observations, and analyzing the results. If a question cannot be answered in this way, then it is not a science question.

Scientists bring their own ideas that can influence their results. They try to think with an open mind. They use creativity when they try to find solutions to problems and answers to questions. Are you ready to think like a scientist? Then let's go!

DCI LS4.A: **Evidence of Common Ancestry and Diversity.** Fossils provide evidence about the types of organisms that lived long ago and also about the nature of their environments. (3-LS4-1)
NS Science Is a Human Endeavor. Science affects everyday life. (3-ESS3-1)

At a fossil dig site, scientists from different backgrounds work as a team to make sense of what they have found.

Wrap It Up!

My Science Notebook

1. **Summarize** What are some characteristics that are valuable in science?

2. **Infer** What are some advantages to having a group of scientists with different backgrounds work together?

15

Practice Science

? **How can you make inferences from fossil fragments?**

Sometimes scientists only find fragments, or pieces, of a fossil, rather than a fossil of a whole organism. The fragment might be a part of a tooth, a shell, or a bone or other structure from the body. Scientists observe these fragments and make inferences about what the organism may have looked like. They also make inferences about the environment the organism lived in and the conditions of that environment. In this investigation, you will make observations of a fossil fragment and then make inferences based on your observations.

Materials

fossil fragment	ruler
colored pencils	hand lens

DCI LS4.A: Evidence of Common Ancestry and Diversity. Fossils provide evidence about the types of organisms that lived long ago and also about the nature of their environments. (3-LS4-1)
CCC Patterns. Similarities and differences in patterns can be used to sort and classify natural phenomena. (3-LS3-1)
NS Scientific Knowledge Is Based on Empirical Evidence. Science findings are based on recognizing patterns. (3-LS1-1), (3-PS2-2)

1 Observe your fossil fragment without the hand lens. Measure and record the dimensions of your fossil fragment.

2 Choose one color of pencil to draw your fossil fragment. Your drawing may be larger or smaller than the actual size of your fossil fragment. If needed, include a key that shows the scale of the drawing, such as half of the real size or twice the real size.

3 Observe your fossil fragment using the hand lens. What can you see with the hand lens that you couldn't see before? Add details to your drawing. Use the same color pencil as in Step 2.

4 Based on your observations, make inferences about what the rest of the organism might have looked like. Use a different colored pencil to draw the rest of the organism.

5 Make further inferences about the environment the organism may have lived in. Draw the environment around your organism.

6 Share your drawings with the rest of the class. Discuss the inferences you made to decide what your organism might have looked like and the environment it lived in.

Wrap It Up! **My Science Notebook**

1. **Explain** How did you use both observation and inference in this activity?

2. **Construct an Argument** Construct an argument about what your organism looked like or characteristics it had based on evidence from your fossil fragment.

3. **Apply** In what ways would a scientist's efforts to create a model from the fossil fragment be like yours? In what ways would they differ?

Andrés Ruzo Geoscientist
National Geographic Explorer

Let's Explore!

In *Nature of Science,* you learned that scientists study the natural world. It is easy to see how plants, animals, clouds, and even boiling rivers are all part of the natural world. But what about machines? Do scientists study human-made objects? Yes! We do! Because how objects interact, how they move, and the patterns they follow are all part of the natural world. Look for patterns of motion in the lessons ahead.

Physical science is the study of nonliving objects and systems. Here are some questions you might answer in *Physical Science*:

- How does a soccer player change the speed and direction of a ball?

- How can you use a magnet to make a paper clip seem to hover in the air?

- How can you make and test your own magnet?

- Why does your hair stand on end when you bring a charged balloon near your head?

Look at the notebook examples for some ideas. Let's check in again later to review what you have learned!

▼ Define and illustrate new science vocabulary and concepts.

▼ Include captions and drawings to explain what you have learned.

Unbalanced Forces

The player adds a force to the ball when he kicks it. The forces are unbalanced. The ball moves.

What I Learned

Charged Objects
I learned that charged objects can attract or repel each other. Objects with opposite charges attract each other.

Fossils
I learned that fossils help us understand what the environment was like long ago. Fish fossils found in a desert show that the whole area was once covered by water.

▶ Include notes with drawings to explain main ideas.

Balanced Forces

The net force on the ball is zero. The forces are balanced. The ball does not move.

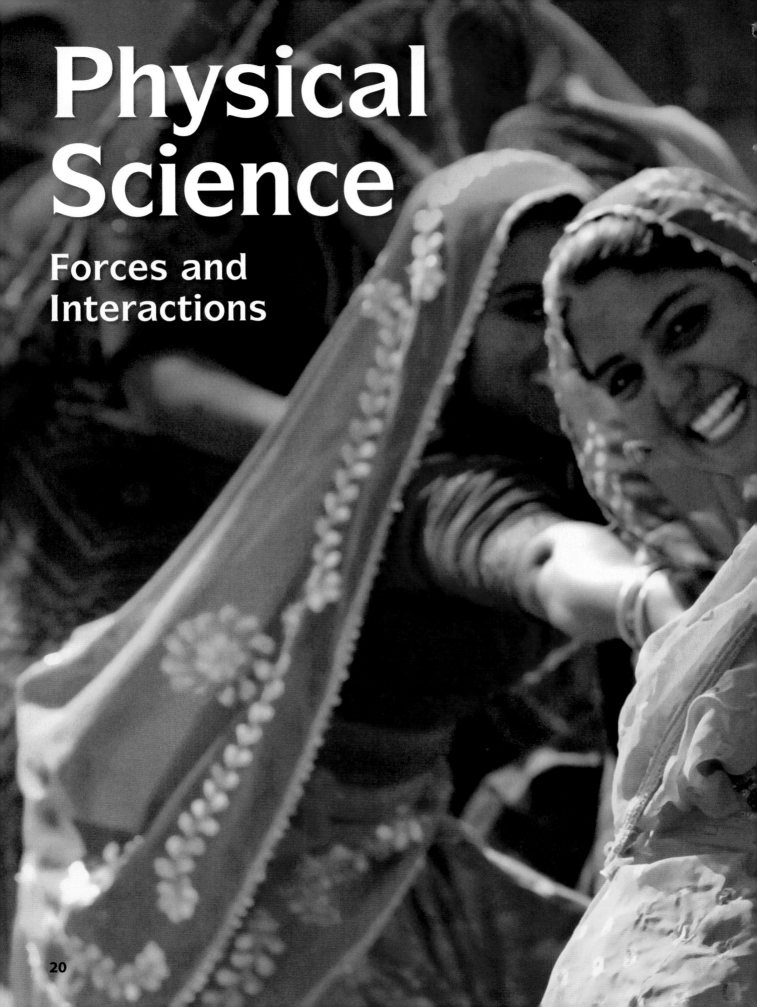

Physical Science

Forces and Interactions

Women pull on a rope during a tug of war match.

Pushes and Pulls

These athletes are in a race across dunes in the Sahara, a desert in North Africa. They use their strength to apply forces in the race. A **force** is a push or a pull. Every time an object is moved, a force is acting on it. You use patterns of forces when you push on the pedals of your bicycle or pull a door closed.

Athletes in this part of the race push and pull a team member in a special wheelchair.

DCI PS2.A: Forces and Motion. Each force acts on one particular object and has both strength and a direction. An object at rest typically has multiple forces acting on it, but they add to give zero net force on the object. Forces that do not sum to zero can cause changes in the object's speed or direction of motion. (3-PS2-1)
CCC Patterns. Patterns of change can be used to make predictions. (3-PS2-2)

Every force has a strength and a direction. In this desert race, team members in front pull the wheeled cart. Those behind it push the cart. All the team members are applying forces in the same forward direction. If the forces are strong enough, the cart will move.

Wrap It Up! 📓 My Science Notebook

1. **Define** What is a force?

2. **Relate** How does the direction of a push relate to the direction that the object moves?

3. **Predict** Suppose more team members pushed and pulled the cart. How might the force the people apply to the cart change?

Balanced Forces

The blackbelt studies four boards. The boards rest on two concrete blocks. When objects are in contact, they exert forces on each other. Gravity pulls down on the boards. The cement blocks push up on them. These **balanced forces** cancel each other out. The **net force,** or overall force, on the boards is zero.

While forces acting on the boards are balanced, the boards do not move.

HYAH! The blackbelt hits the top board! The hand striking the board applies another force to it. The added downward force causes all the boards to move. The boards break and their pieces scatter.

When the blackbelt's hand strikes the top board, the forces acting on the board are no longer balanced.

DCI PS2.A: Forces and Motion. Each force acts on one particular object and has both strength and a direction. An object at rest typically has multiple forces acting on it, but they add to give zero net force on the object. Forces that do not sum to zero can cause changes in the object's speed or direction of motion. (3-PS2-1)
DCI PS2.B: Types of Interactions. Objects in contact exert forces on each other. (3-PS2-1)
CCC Cause and Effect. Cause and effect relationships are routinely identified. (3-PS2-1)

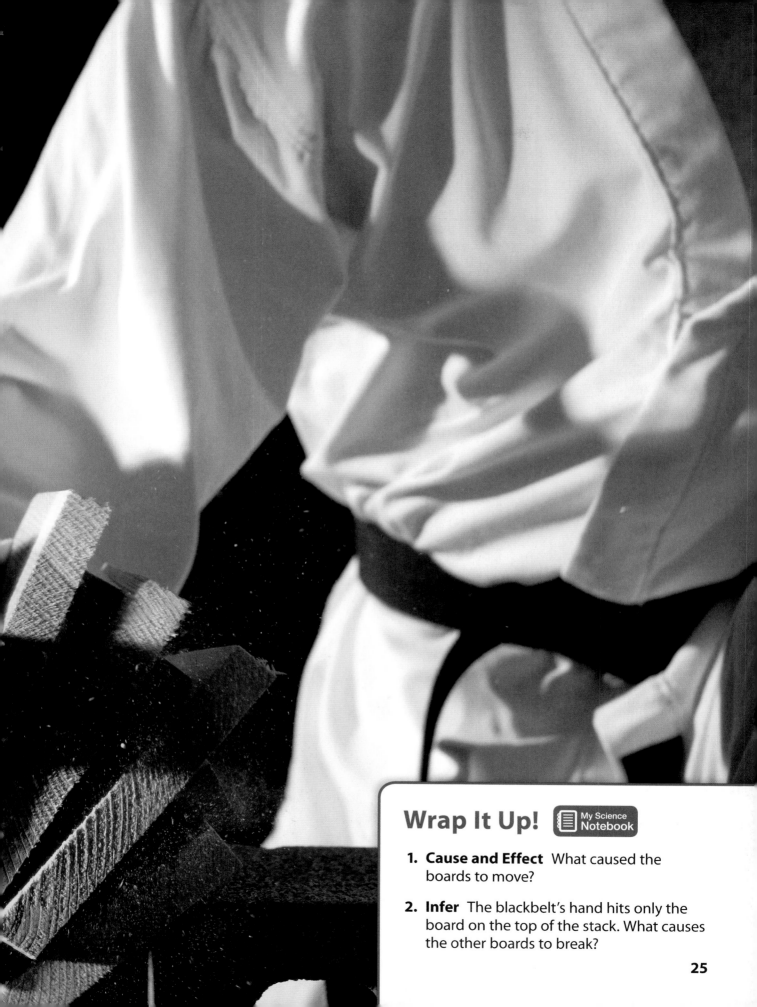

Wrap It Up! 🗒️ My Science Notebook

1. **Cause and Effect** What caused the boards to move?

2. **Infer** The blackbelt's hand hits only the board on the top of the stack. What causes the other boards to break?

25

Unbalanced Forces

A game of tug-of-war can be fun! Both team's members hold the rope tightly in their hands. The rope stays in one place as both teams pull with all their might. When the two teams pull in opposite directions with the same force, the net force is zero. The forces on the rope are balanced.

DCI PS2.A: Forces and Motion. Each force acts on one particular object and has both strength and a direction. An object at rest typically has multiple forces acting on it, but they add to give zero net force on the object. Forces that do not sum to zero can cause changes in the object's speed or direction of motion. (3-PS2-1)
DCI PS2.B: Types of Interactions. Objects in contact exert forces on each other. (3-PS2-1)

Now the women on the right pull with more force than the men on the left. The rope is moved to the right, the direction in which the women are pulling. The forces on the rope are unbalanced. **Unbalanced forces** cause an object to move.

When forces are unbalanced, they no longer add up to zero. In this case, the net force is to the right, and the rope is moved to the right.

Villagers in China play tug of war. The team that applies the stronger pulling force will win.

Wrap It Up! 📓 My Science Notebook

1. **Define** What is an unbalanced force?

2. **Infer** Suppose the rope is moved to the left. In which direction is the net force?

3. **Apply** Describe a situation in which forces are balanced and a situation in which forces are unbalanced.

Building Bridges

Problem

How can engineers design a safe bridge?

On May 23, 2013, part of the Skagit River Bridge in the state of Washington collapsed into the river below. The bridge had recently been judged to be in good condition. But a truck carrying an oversized load struck part of a bridge support. This damage caused the collapse. Luckily, there were no major injuries. Engineers study conditions that cause bridge failure, so they can design safer bridges.

Engineers are always working to improve bridge design. They have done this in the past, too. The Golden Gate Bridge in San Francisco, California, opened in 1937. Engineers faced many challenges to design a safe bridge. Tides, fog, wind, and salty air made the bridge difficult to build and maintain. In addition, earthquakes are a threat in this area. The bridge would need to balance the downward force provided by all the vehicles that use the bridge.

DCI PS2.A: Forces and Motion. Each force acts on one particular object and has both strength and a direction. An object at rest typically has multiple forces acting on it, but they add to give zero net force on the object. Forces that do not sum to zero can cause changes in the object's speed or direction of motion. (Boundary: Qualitative and conceptual, but not quantitative addition of forces are used at this level.) (3-PS2-1)
CETS Interdependence of Science, Engineering, and Technology. Scientific discoveries about the natural world can often lead to new and improved technologies, which are developed through the engineering design process. (3-PS2-4)
CETS Influence of Engineering, Technology, and Science on Society and the Natural World. People's needs and wants change over time, as do their demands for new and improved technologies. (3–5-ETS1-1) • Engineers improve existing technologies or develop new ones to increase their benefits, decrease known risks, and meet societal demands. (3–5-ETS-2)

The Skagit River Bridge collapse was caused by accidental damage to one of the supporting structures. The supporting structure was needed to balance the downward forces on the bridge.

Solution

To keep a bridge from collapsing, the upward force of cables supporting the roadbed and vehicles balance the downward force of gravity. Engineers designed a suspension bridge with strong towers and cables. They tested models of the bridge.

The steel towers that support the cables are very important. One of the towers is anchored deep underwater in a concrete mass about the size of a football field. The cables were so huge and strong that they had to be made at the site. Thin steel wires were bound together to make the cables.

A suspension bridge has a very flexible design. The Golden Gate Bridge allows for movement caused by winds or earthquakes. In steady winds, the bridge can move side to side up to about 8 meters (27 feet) without damage. Parts of the roadbed can move up and down more than 4 meters (about 16 feet)!

More than two billion vehicles have crossed the bridge since it was opened. The bridge is painted regularly with an orange-colored paint that protects the metal parts from the weather. Iron and steel workers repair damaged parts. By making sure forces are safely balanced, the Golden Gate Bridge should be around for billions more vehicles to cross.

Workers built the Golden Gate Bridge section by section in the 1930s. It has withstood the forces of heavy vehicles, wind, and earthquakes.

The roadway of a suspension bridge is held up by huge cables that hang over the towers. The cables are anchored deep in concrete at both ends. Smaller cables connect the roadway to the main cables.

Changing Direction

Athletes use forces in games of all types. This photo shows a type of kick volleyball called sepak takraw that is often played in Southeast Asia.

A skilled player can control every motion of the ball. The player uses forces to start the ball moving, changing its direction and speed.

When the player's foot contacts the ball, his foot exerts a force on the ball and the ball exerts a force on his foot. The player's kick changes the direction of the ball.

DCI PS2.A: **Forces and Motion.** Each force acts on one particular object and has both strength and a direction. An object at rest typically has multiple forces acting on it, but they add to give zero net force on the object. Forces that do not sum to zero can cause changes in the object's speed or direction of motion. (3-PS2-1)

DCI PS2.B: **Types of Interactions.** Objects in contact exert forces on each other. (3-PS2-1)

A strong force will cause the ball to move faster than a weak force. He uses just the right force in the right direction to move the ball over the net.

Think of the force you would need to move this ball over the net. Would you need a push or a pull? How strong of a force would you need? In which direction would you apply the force?

Change of Motion

1 Apply a gentle force to a ball by tapping it with a pencil. How did you change its motion?

2 How else can you change the motion of the ball? Use the pencil to change the motion of a ball in different ways. Can you make the ball move faster? Slower? To the left or right? Can you make it stop?

Wrap It Up!

1. **Recall** Are the forces on the ball in the photo balanced or unbalanced? How do you know?

2. **Generalize** In what ways can forces change an object's motion?

Plan and Conduct an Investigation

These players use balanced and unbalanced forces to control the motion of a ball. They apply a force to the ball when they kick it in the direction they want it to go. Now you will plan an investigation to test the effects of forces on objects.

1. **Ask a question.** My Science Notebook

   How do balanced and unbalanced forces determine an object's motion?

2. **Plan and conduct an investigation.**

   Build a launcher. Wrap a rubber band around a sturdy folder. Crumple a small piece of paper and place it in front of the rubber band. Pull back the rubber band and let go. Observe the effect of the force on the paper ball.

   Look at the materials available. Plan an investigation to test the effects of different strengths of force on the ball. Make your test fair. Measure the strength of force and measure each result in the same way. Write the steps of your plan in your notebook.

   Use these questions to help you plan:

   - How will you measure the strength of force?
   - How will you vary the strength of force?
   - How will you measure the motion of the ball?
   - How many trials will you run?

   Conduct your investigation. Record your observations in your notebook.

PE 3-PS2-1. Plan and conduct an investigation to provide evidence of the effects of balanced and unbalanced forces on the motion of an object.

3. **Analyze your results.**
   Review your observations. How did the force of the rubber band affect the ball's motion? When did you observe unbalanced forces? When did you observe balanced forces?

4. **Share your results.**
   Discuss your results with your group. Explain how forces affect the object's motion when it was moving and when it was not moving. Explain when forces were balanced and when they were unbalanced.

5. **Explain your findings.**
   Organize your results. Make a graph to show which force made the ball move the farthest. Present your findings to the class.

Explore on Your Own
Plan another fair test. This time, change the direction of force instead of the amount of force. Carry out your investigation.

Patterns of Motion

Some motion is hard to predict. For instance, it is hard to know exactly where a leaf blowing in the wind might land. But some motion follows a pattern.

When motion follows a pattern, it is easier to predict future motion. **Regular motion** has a pattern of repeating over and over. A swing has regular motion. Swings move back and forth, over and over.

DCI PS2.A: Forces and Motion. The patterns of an object's motion in various situations can be observed and measured; when that past motion exhibits a regular pattern, future motion can be predicted from it. (3-PS2-2)
CCC Patterns. Patterns of change can be used to make predictions. (3-PS2-2)

The motion of a swing follows a predictable pattern.

Wrap It Up! My Science Notebook

1. **Contrast** How does the motion of a falling leaf differ from the motion of a swing?

2. **Predict** Look at the person in the photo. Describe how she will move next.

Motion

? **How can you predict a marble's motion?**

Some objects' motion can be observed and measured.
When you observe a pattern, you can predict what will
likely happen next. In this investigation, you will
observe and predict the motion of a marble.

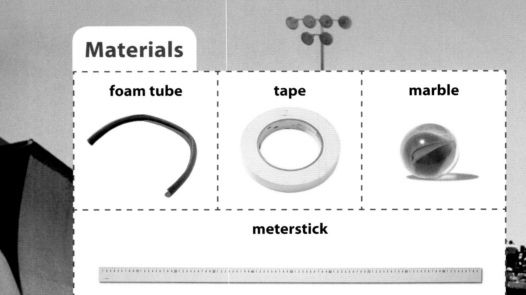

Materials

foam tube	tape	marble
meterstick		

When the skateboarder rolls up
one side of the halfpipe, what
happens next?

DCI PS2.A: Forces and Motion. The patterns of an object's motion in various situations can be observed and measured; when that past motion exhibits a regular pattern, future motion can be predicted from it. (3-PS2-2)
CCC Patterns. Patterns of change can be used to make predictions. (3-PS2-2)

1 Set up two chairs as shown. Tape each end of the ramp to the back of a chair. The tube should touch the floor in between the chairs. Use the meterstick to measure the height of the ramp. Record the measurement in your notebook. Place the marble at the top of the ramp.

2 Let the marble go. Observe how far the marble moves up the other side of the ramp. Put your finger in that spot and have a partner measure the height. Repeat your test three times. Record your observations.

3 Choose a different height to drop the marble from. Predict how far the marble will move up the other side of the ramp. Release the marble. Repeat three times and record your observations.

4 Repeat step 3 from two different heights.

Wrap It Up!

1. **Predict** Did your results support your predictions? Explain.

2. **Interpret** What pattern do you notice in your data?

Math Girl

In Russia and elsewhere in the 1800s, education systems supported males in math and science. Girls and women who wanted to study math and science had to fight for it.

Sofia fought to become a professor at the University of Stockholm in Sweden. She was one of the first women in northern Europe to become a professor.

DCI PS2.A: Forces and Motion. The patterns of an object's motion in various situations can be observed and measured; when that past motion exhibits a regular pattern, future motion can be predicted from it. (3-PS2-2)
NS Scientific Investigations Use a Variety of Methods. Science investigations use a variety of methods, tools, and techniques. (3-PS2-1)

Sofia's struggle has helped girls and women achieve their dreams.

Sofia Kovalevskaya was born in Russia in 1850. As an adult, she made major contributions to science. That was not easy, though. Math is important in science. It was very hard for girls to pursue math in Europe then. But she was determined.

Young Sofia showed a talent for math. So, her parents hired a tutor. But that was as far as she could go in Russia. Women could not attend a university. She needed written permission from her father or a husband. Her father refused, so she married to get permission. She then moved to Germany. There, she could attend a university. Her problems did not end. Sofia's professors would not grade her work. After studying for two years, she hired private tutors.

A professor finally helped Sofia to earn a doctorate in math. But this still wasn't easy. She had not gone to the required classes or passed the needed exams. She wrote three papers to get her degree. One was about the patterns of Saturn's rings. Sofia was the first woman in Europe to earn a doctorate degree. After that, she became a professor.

Sofia fought for equal rights for women. She showed that women could do math and science and do them well.

Wrap It Up!

1. **Infer** Why do you think math is important in science?

2. **Describe** What characteristics helped Sofia achieve her dreams?

Make Observations

A trapeze is like a playground swing. The trapeze artist uses the pattern of the trapeze's swinging motion to plan and perform tricks. You can observe and predict the motion of a swinging object, too.

1. **Ask a question.**

   How can you use your observations to predict the future motion of a swinging object?

2. **Plan and conduct an investigation.**

   Tie a metal washer to one end of a string. Tape the free end of the string to the edge of a table. The washer should be close to the floor but not touching it. You have made a pendulum!

   Swing the pendulum to see how it moves.

   Plan an investigation to determine how you can make the pendulum move faster, higher, or longer. Record your plan in diagrams or in words. Predict what you think will happen when you make each change. Then carry out your plan. Record your observations as you carry out each test.

3. **Analyze and interpret data.**

   Examine your data. What were your results? Can you use your data to predict the future motion of the pendulum? Are there things you could do differently to improve your results? Revise your plan and retest. Record your observations.

PE 3-PS2-2. Make observations and/or measurements of an object's motion to provide evidence that a pattern can be used to predict future motion.
NS Scientific Knowledge is Based on Empirical Evidence. Science findings are based on recognizing patterns. (3-PS2-2)

4. **Share your results.**

   Explain your results to a partner. How did you collect data? What pattern did you observe? How can patterns of motion be used to make predictions? Have your partner give you feedback. If you need to, revise your method and test again.

5. **Explain your findings.**

   After you are satisfied with your results, present your findings to the class. Demonstrate and explain how you can use your observations to predict the future motion of the pendulum.

A swinging trapeze repeats a pattern of motion.

Magnets

Have you ever dipped a magnet into a pile of paper clips? If you have, you saw that the paper clips stuck to the magnet. That's because **magnets** pull on certain kinds of metals. Magnets also pull on, or **attract,** other magnets. The pull a magnet exerts is called **magnetic force.** A magnet doesn't have to touch a paper clip to cause an effect on it. The magnet just has to be close enough to the clip for the magnetic force to affect it.

The places on a magnet where the pull is strongest are called **poles.** There are two kinds of poles—north (N) and south (S). The north pole of one magnet attracts the south pole of another magnet. Two north poles or two south poles **repel,** or push each other apart.

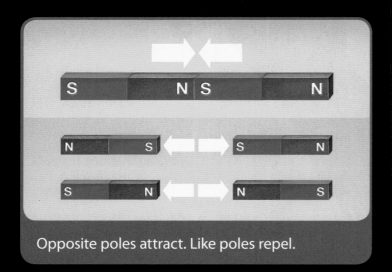

Opposite poles attract. Like poles repel.

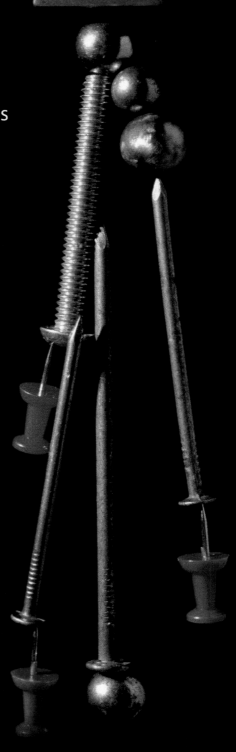

The magnetic force of this magnet is strong enough to keep these objects from falling.

DCI PS2.B: Types of Interactions. Electric and magnetic forces between a pair of objects do not require that the objects be in contact. The sizes of the forces in each situation depend on the properties of the objects and their distances apart and, for forces between two magnets, on their orientation relative to each other. (3-PS2-3), (3-PS2-4)
CCC Cause and Effect. Cause and effect relationships are routinely identified. (3-PS2-1)

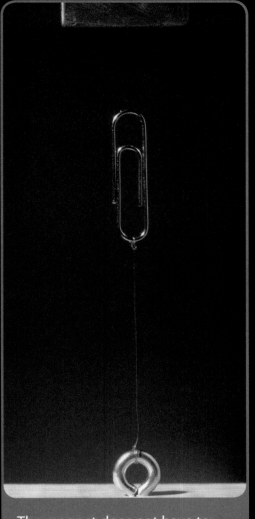

The magnet does not have to touch the paper clip to attract it.

The pieces that make up this sculpture are magnets. Magnetic force holds the sculpture together.

Wrap It Up! My Science Notebook

1. **Define** What does *attract* mean? What does *repel* mean?

2. **Contrast** How is the force applied in the kick of a soccer ball different from the force a magnet exerts on a paper clip?

3. **Explain** Tell what the phrase "opposites attract" means about magnets.

Magnetic Force

? **How do magnets exert force?**

Some magnets can pull harder than others. Remember, every force has a strength and direction. The strength and direction of a magnet's force can depend on the magnet's size, materials the magnet is made of, how far away the magnet is, and how the magnet is turned. In this investigation, you'll explore magnetic forces.

Materials

ruler	2 bar magnets

paper clips	1 small bar magnet

DCI PS2.B: **Types of Interactions.** Electric and magnetic forces between a pair of objects do not require that the objects be in contact. The sizes of the forces in each situation depend on the properties of the objects and their distances apart and, for forces between two magnets, on their orientation relative to each other. (3-PS2-3), (3-PS2-4)
CCC Cause and Effect. Cause and effect relationships are routinely identified, tested, and used to explain change. (3-PS2-3)

1 Place the 2 large bar magnets 10 cm apart with their north poles facing each other. Slowly push the magnet on the right toward the magnet on the left. Record your observations.

2 Repeat step 1 with a north pole facing a south pole and then again with a south pole facing a south pole. Record your observations.

3 Arrange a paper clip and one of the large bar magnets along the ruler as shown. Predict how close the magnet will get before the magnetic force affects the clip. Record your prediction, and then carry out your test. Record your observations.

4 Slowly move the north pole of 1 large bar magnet near the paper clips. Repeat with the south pole. Record your observations. Repeat using the smaller bar magnet. Record your observations.

Magnets attracts shavings of magnetic metal.

Wrap It Up! My Science Notebook

1. **Describe** Identify evidence from your investigation that magnets can exert forces without touching.

2. **Compare and Contrast** How are the large and small magnets alike? How are they different?

3. **Cause and Effect** What would happen if you brought the north pole of a bar magnet toward the north pole of another magnet that was attached to a toy car?

Electromagnets

? **How can you test the strength of an electromagnet?**

Some magnets use electricity to produce a very strong magnetic force. These magnets are called electromagnets. Electromagnets are handy at scrap yards because they are very strong, and they can be turned on and off. In this investigation, you'll build and test an electromagnet.

Materials

bolt	wire	battery

battery holder	**paper clips**	

DCI PS2.B: Types of Interactions. Electric and magnetic forces between a pair of objects do not require that the objects be in contact. The sizes of the forces in each situation depend on the properties of the objects and their distances apart and, for forces between two magnets, on their orientation relative to each other. (3-PS2-3), (3-PS2-4)

CCC Cause and Effect. Cause and effect relationships are routinely identified, tested, and used to explain change. (3-PS2-3)

1 Leave about 10 cm of wire loose at one end and wrap the wire around the bolt 15 times. Try not to overlap the wire as you wrap it around the bolt.

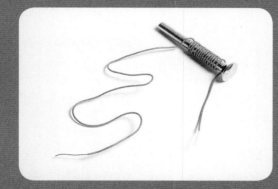

2 Attach each end of the wire to the battery holder by connecting it to the metal pieces on each side. Place the battery in the holder.

3 Test your electromagnet. Bring the end of the bolt near a pile of paper clips. Count the number of paper clips the electromagnet can hold. Record your observations.

4 Remove the battery. Predict what will happen if you wrap the wire around the bolt 25 times. Try it! Record your observations. Repeat with 35 wraps.

Electromagnets can be very powerful.

Wrap It Up!

1. **Explain** How did you measure the strength of the electromagnet's force? When was it weakest? Strongest?

2. **Generalize** Can an electromagnet exert a force without touching an object? Explain.

3. **Cause and Effect** What might happen if you had a longer bolt and wrapped the wire around 50 times?

STEM

ENGINEERING PROJECT

Design a Crane

Lifting and moving heavy loads is a problem in some factories. If the loads are metal objects, electromagnetic cranes can solve the problem. They can lift extremely heavy loads—as much as eight cars of weight! And they can be turned off to drop the load where needed.

A toy company wants to make a toy crane that can pick up metal objects. They need your team of engineers. You have observed how electromagnets work. You have even built your own electromagnet. Now you will use what you have learned to design a prototype of the new toy.

The Challenge

Your challenge is to design and build a prototype of a toy crane. Your prototype must:

- use an electromagnet
- pick up and drop off three paper clips
- lift the paper clips to three centimeters or more

PE **3–5-ETS1-1.** Define a simple design problem reflecting a need or a want that includes specified criteria for success and constraints on materials, time, or cost.
PE **3-PS2-4.** Define a simple design problem that can be solved by applying scientific ideas about magnets.

An electromagnetic crane uses magnets and electricity to get the job done.

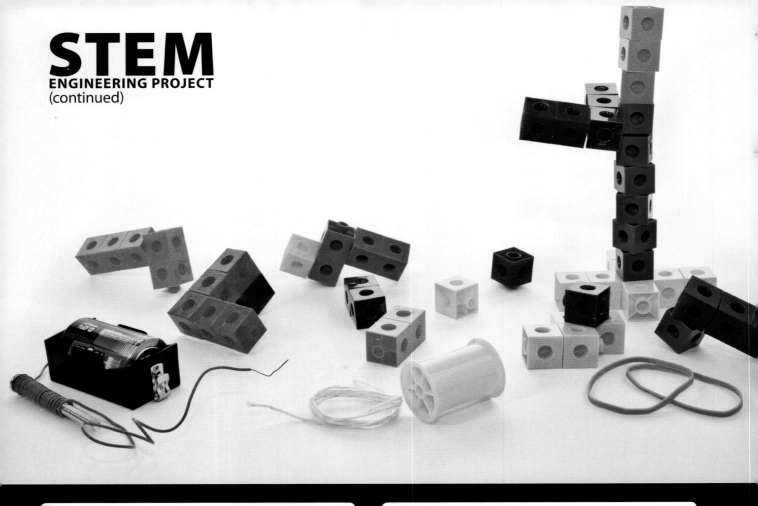

STEM
ENGINEERING PROJECT
(continued)

1 Define the problem. My Science Notebook

Think about the problem you are solving. Describe the need that your team will try to fill. Write the problem in your science notebook.

What exactly does your prototype need to do? Those things are the criteria of the problem. Criteria tell you if your design is successful. Look back at the Challenge box to find your design criteria.

Your teacher will give you the materials for your prototype. You cannot use any other materials. Limits on the materials you can use are constraints on your design. Other constraints include the amount of time and space you have to build your solution.

List the criteria and constraints in your science notebook.

2 Find a solution.

Your team will use the electromagnet you built earlier. You will use connecting blocks to make the body of your crane. Think about these questions:

- How will you attach your electromagnet?
- How will you lift the paper clips?

Draw your design, and share it with your team. Discuss all of your designs. Choose the best design to build and test.

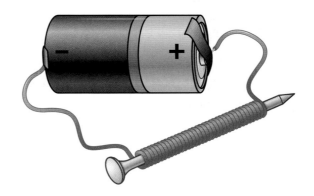

Think about how your crane will support the weight of the bolt. You may consider attaching the battery case to the base of the crane.

3 Test your solution.

Build your electromagnet, or use the one you have built during an earlier activity. Have each member of the team try picking up and dropping a paper clip using the electromagnet.

Follow your design plan to build the rest of the crane. Is the crane strong enough to hold the electromagnet? Try moving the electromagnet up and down. Adjust your prototype to make it work. Make notes to show what you change.

Test your prototype. Can it pick up three paper clips? Use the ruler to measure how high it lifts the paper clips. Write your observations in your notebook.

Discuss the results of the test with your team. Did your prototype meet the criteria?

4 Refine or change your solution.

Talk about ways to improve your prototype of a toy crane. Think about what you know about magnets and magnetic force. Use your ideas to make changes. Record all your changes in your notebook.

How will you know if your changes improved your crane? Test your prototype again. Did your changes make a difference? What observations tell you whether it made a difference or not?

Present your prototype to the class. Describe the results of your tests. Answer questions about your design. Ask questions about the other teams' prototypes.

How could you further improve your design? Record your ideas in your notebook.

Electric Forces

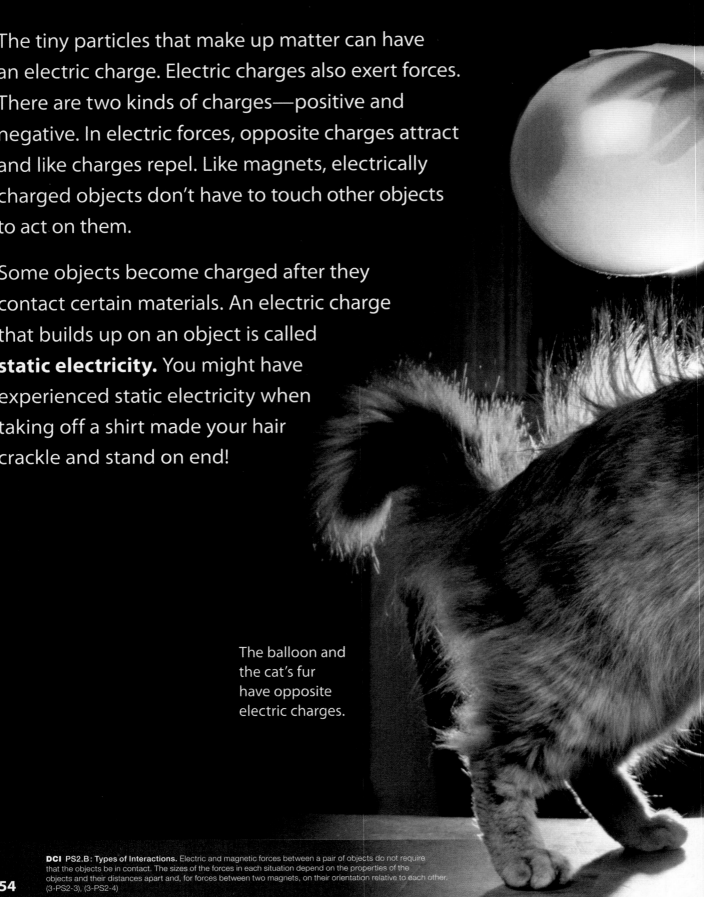

The tiny particles that make up matter can have an electric charge. Electric charges also exert forces. There are two kinds of charges—positive and negative. In electric forces, opposite charges attract and like charges repel. Like magnets, electrically charged objects don't have to touch other objects to act on them.

Some objects become charged after they contact certain materials. An electric charge that builds up on an object is called **static electricity.** You might have experienced static electricity when taking off a shirt made your hair crackle and stand on end!

The balloon and the cat's fur have opposite electric charges.

DCI PS2.B: Types of Interactions. Electric and magnetic forces between a pair of objects do not require that the objects be in contact. The sizes of the forces in each situation depend on the properties of the objects and their distances apart and, for forces between two magnets, on their orientation relative to each other. (3-PS2-3), (3-PS2-4)

SCIENCE in a SNAP

Effects of Electric Charge

1 Cut tissue paper into small pieces. Put the pieces in a pile.

2 Move a balloon toward the pieces of tissue paper. Record your observations.

3 Rub one side of the balloon with a wool cloth. Predict what will happen when you move the rubbed balloon toward the tissue paper. Then try it!

Wrap It Up!

1. **Compare** How are the forces exerted by electric charges similar to the forces exerted by magnets?

2. **Identify** Where have you seen the effects of static electricity at home or at school?

Electric Forces

? **How can you observe the effects of electric forces?**

You've investigated how magnetic forces interact. Now you can investigate the interaction of electric forces.

Materials

2 balloons	2 strings

tape	**wool cloth**

The foam packing pieces are not sticky like tape. It is static electricity that makes them cling to the woman's hands.

DCI PS2.B: Types of Interactions. Electric and magnetic forces between a pair of objects do not require that the objects be in contact. The sizes of the forces in each situation depend on the properties of the objects and their distances apart and, for forces between two magnets, on their orientation relative to each other. (3-PS2-3), (3-PS2-4)

1 Tie a piece of string to each balloon. Tape the ends of both balloon strings to the edge of a table. Allow the balloons to hang freely, about 5 cm apart.

2 Move the balloons toward each other, and then let them go. Record your observations.

3 Place one balloon on the table. Rub the balloon with the wool cloth. Be sure to rub all parts of the balloon. Predict what will happen when the balloons again hang freely. Then try it. Record your predictions and observations.

4 Now rub both balloons with the wool cloth. Be sure to rub all parts of the balloons. Predict what will happen when the balloons again hang freely. Then try it. Record your predictions and observations.

Wrap It Up!

1. **Explain** Why did you observe the hanging balloons before you rubbed them?

2. **Describe** In steps 3 and 4, what happened when the balloons were hanging freely? What did you do to cause this difference?

3. **Infer** What can you infer about the charges on the balloons in step 3? In step 4?

Determine Cause and Effect Relationships

You have observed magnetic and electric forces at work. You have also seen that forces have different strengths and directions. The magnet in the photo is so strong it can exert force through a hand! Now it's your turn to investigate magnetic or electric forces.

1. **Ask a question.**

   Choose one of these questions to test:
   - How can I measure the strength of a magnet?
   - How can I determine the direction of a magnet's force?
   - How can I measure the strength of a static electric charge?
   - How can I determine the direction of force of a charged object?

2. **Plan and conduct an investigation.**
   Copy the question you chose into your notebook. Look at the materials available to you. Plan an investigation to test your question. List and gather the materials you will need. Write the steps of your investigation. Then carry them out.

3. **Analyze and interpret data.**
   Examine your data. What did your results show? Were you able to measure the strength of a magnet or an electric charge, or the direction of a magnet's force? Did your investigation show the relationship between a cause and its effect? If not, how could you change your investigation so that it shows such a relationship and answers your question?

PE 3-PS2-3. Ask questions to determine cause and effect relationships of electric or magnetic interactions between two objects not in contact with each other.

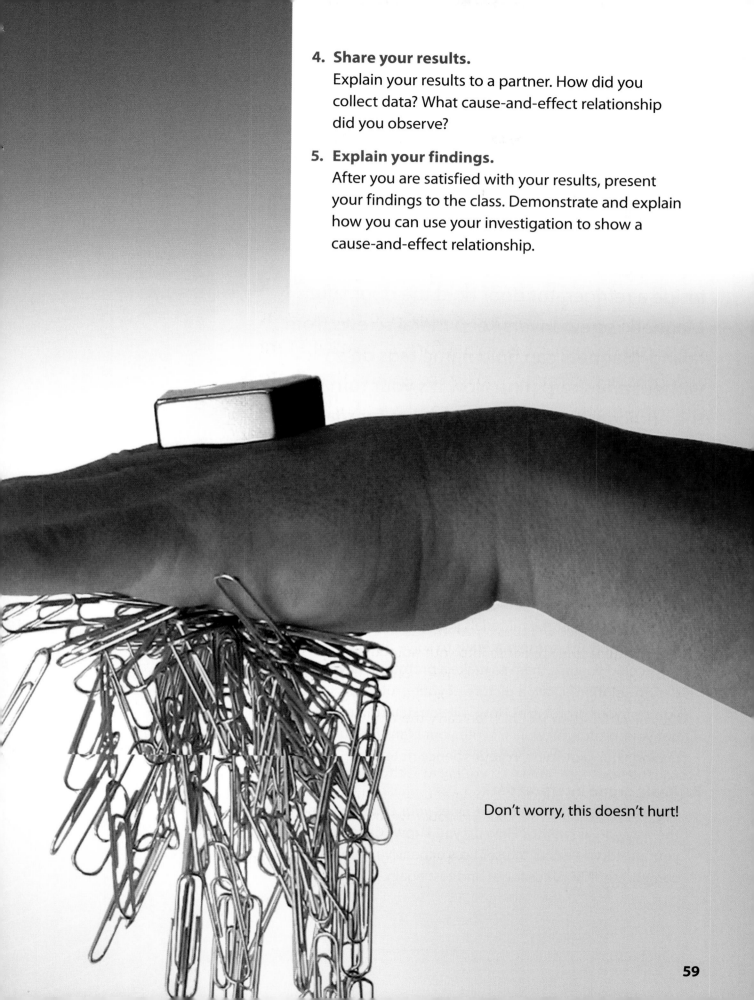

4. **Share your results.**
   Explain your results to a partner. How did you collect data? What cause-and-effect relationship did you observe?

5. **Explain your findings.**
   After you are satisfied with your results, present your findings to the class. Demonstrate and explain how you can use your investigation to show a cause-and-effect relationship.

Don't worry, this doesn't hurt!

Define and Solve a Design Problem

Magnets have many practical uses. Magnets inside a refrigerator door hold the door shut. Magnetic screwdrivers keep metal screws from falling. Magnets can hold name tags on shirts without any sharp pins. Now it's your turn to be the engineer and design a new use for magnets.

1. **Ask a question.** 📓 **My Science Notebook**

   How can you use magnets to solve a problem?

2. **Plan and carry out an investigation.**

   Think of a problem that can be solved using magnets. For example, could you use magnets to keep a mailbox shut? Could you use magnets to keep two toy cars on a track from crashing together? Write the problem in your notebook. Look at the materials your teacher provides. You may only use these materials to solve the problem. How could you use them to make a prototype of your design solution? Draw a picture of your prototype. How will you test your prototype? How will you determine whether or not your prototype works? Write your plans for building and testing your prototype in your science notebook.

3. **Analyze and interpret data.**

   Make and test your prototype. Does your solution work the way you want it to? How do you know? Can you make your prototype easier to use? Less expensive to make? Less complicated? Make changes and test again.

This ball contains a sensor that detects a magnetic field at the opening of the goal.

PE 3-PS2-4. Define a simple design problem that can be solved by applying scientific ideas about magnets.
PE 3-5-ETS1-1. Define a simple design problem reflecting a need or a want that includes specified criteria for success and constraints on materials, time, or cost.

4. **Construct an explanation from evidence.**
Explain your prototype to a partner. How does it use magnetic forces to solve a problem? Does it use magnets to attract or repel an object?

5. **Communicate information.**
Share your results with the class. Now write and investigate your own question.

When the ball crosses the magnetic field, it sends a signal to the referee's watch. Goal!

Roller Coaster Designer

As pushes and pulls move roller coasters, work is being done. But it is fun work! Roller coaster designer Cynthia Emerick Whitson takes advantage of different forces to design roller coasters.

NGL Science What is your job?

Cynthia Emerick Whitson I oversee the design and installation of roller coasters. I figure out how to do it in a way that's safe and not too expensive. My team and I solve engineering problems every day.

NGL Science What did you do in school to learn how to do your job?

Cynthia Emerick Whitson I always found math and science interesting. In college, I learned about materials and how to put them together to make things. I also learned about what happens when materials break down. That helps me understand how to make rides safe.

Roller coasters use motors to move the riders to the top of the first hill. From there, gravity takes over. Forces move roller coasters in all directions.

DCI PS2.A: Forces and Motion. The patterns of an object's motion in various situations can be observed and measured; when that past motion exhibits a regular pattern, future motion can be predicted from it. (3-PS2-2)
NS Scientific Investigations Use a Variety of Methods. Science investigations use a variety of methods, tools, and techniques. (3-PS2-1)

Cynthia Emerick Whitson studied material science and engineering to become a roller coaster designer. She works with teams of people over many months to solve the problems of making rides both thrilling and safe.

Check In

Congratulations! You have completed *Physical Science*. Let's reflect on what you have learned. Here is a checklist to help you judge your progress. Look through your science notebook to find examples of each item in the list.

What could you do better? Write it on a separate page in your science notebook.

▼ Read each item in this list. Ask yourself if you think you did a good job of it.

For each item, select the choice that is true for you: A. Yes **B.** Not Yet

- I defined and drew pictures of science vocabulary, science concepts, and main ideas.
- I labeled drawings. I included captions and notes to explain ideas.
- I collected objects such as photos and magazine or newspaper clippings.
- I used tables, charts, or graphs to record observations and data in investigations.
- I recorded evidence for explanations and conclusions in investigations.
- I described how scientists and engineers answer questions and solve problems.
- I asked new questions.
- I did something else. (Tell about it.)

Reflect on Your Learning

1. Choose one conclusion you made. In what ways did your conclusions reflect the data and what you already knew?

2. Describe a science skill in which you've made progress since beginning your notebook.

Andrés needs good notes to remember everything he observes in the rain forest.

Andrés uses a special tool, his thermal camera, to safely measure temperatures from a distance at the Boiling River of the Amazon.

Andrés Ruzo Geoscientist
National Geographic Explorer

Let's Explore!

There are many ways to do science. Sometimes we use models. I use computer models to study rocks. Earlier, you learned about scientific models. Scientific models are not for decoration. They must be useful! Models can be used to explain how a system or a cycle works. Models can also be used to make predictions. Look for ways that scientists use models in the lessons ahead.

Life science is the study of living things and their environments. Here are some questions you might answer as you read *Life Science:*

- Why does a bald eagle have a hooked beak?
- How does hibernation help a dormouse survive through cold winter temperatures?
- How can living in a group help meerkats survive?
- Why would a fish fossil be found in a desert?
- What does a marine ecologist do?

Look at the notebook examples for some ideas. As you read, think of your own questions and look for answers. Then let's check in again to review what you have learned!

Life Cycle of a Frog

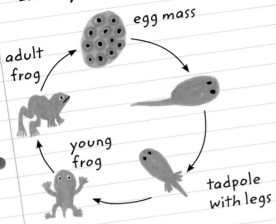

egg mass

adult frog

young frog

tadpole with legs

Main idea: An adult female frog lays eggs. Tadpoles hatch from the eggs. The tadpole develops into an adult frog. The adult frog lays eggs.

My New Questions

1. How do different insects survive through cold winter temperatures?

2. How do frogs and fish that live in a pond ecosystem survive through cold winter temperatures?

3. What kind of plants could I put in my yard to attract wildlife, such as birds and butterflies?

4. What kind of fossils, if any, can be found in my area?

▶ Organize science drawings and concepts in tables and charts.

Living and Nonliving Parts of an Ecosystem

Living	Nonliving
tree	rock
ladybug	cloud
	water

Life Science

Inheritance and Variation of Traits: Life Cycles and Traits

Interdependent Relationships in Ecosystems

The Beaufort's crocodilefish blends in with its surroundings.

Life Cycles

A baby orangutan is at the beginning of its life. It looks similar to its mother, but it is also different. As the young orangutan grows, its body will change, or develop. In 10 to 15 years, the orangutan will become an adult. Then it will have young, or **reproduce.** After many years, the orangutan will grow old and die.

A **life cycle** is a series of changes a living thing goes through during its lifetime. Each kind of organism has its own particular life cycle. Living things begin life, and then grow and develop. Many reproduce, and all finally die. Individuals do not live forever. Reproduction lets each kind of plant and animal continue to live on Earth.

A baby orangutan depends on its mother after birth. The baby and its mother show two stages in the life cycle of an orangutan.

DCI LS1.B: Growth and Development of Organisms. Reproduction is essential to the continued existence of every kind of organism. Plants and animals have unique and diverse life cycles. (3-LS1-1)
CCC Patterns. Similarities and differences in patterns can be used to sort and classify natural phenomena. (3-LS3-1)

Wrap It Up! 📓 My Science Notebook

1. **Define** What is a life cycle?

2. **Identify Patterns** How are a baby orangutan and its mother alike? How are they different?

3. **Apply** What are the stages in the life cycle of a cat?

71

Life Cycle of a Jalapeño Pepper Plant

Have you ever eaten a jalapeño pepper? If so, you know that they are spicy hot! Those peppers are the fruits of jalapeño pepper plants.

Jalapeño pepper plants are flowering plants. Most flowering plants go through similar stages of life. Look at the diagram of the life cycle of the jalapeño pepper plant. Read about each stage.

Jalapeño peppers contain chemicals that make them taste HOT!

 DCI LS1.B: Growth and Development of Organisms. Reproduction is essential to the continued existence of every kind of organism. Plants and animals have unique and diverse life cycles. (3-LS1-1)

SEED Each fruit holds many seeds.

ADULT PLANT Flowers can grow on an adult pepper plant. The flowers may produce fruit—the jalapeño peppers.

SEEDLING Given moisture and the right temperature, a pepper seed can grow into a seedling.

YOUNG PLANT The seedling grows into a young pepper plant.

Wrap It Up! 📓 My Science Notebook

1. **Identify** Which parts of a pepper plant produce fruit?

2. **Sequence** Put these life cycle stages in order: young plant, seedling, seed, adult plant. Start with a seed.

3. **Analyze** In which stage of its life cycle does a pepper plant reproduce?

Life Cycle of a Ladybug

Ladybugs are small, spotted, oval-shaped insects. The ladybug looks different during each stage of its life cycle. Trace the diagram of the ladybug life cycle as you read about each stage.

Ladybugs eat many different types of insects that attack crops, such as aphids and mealybugs.

DCI LS1.B: Growth and Development of Organisms. Reproduction is essential to the continued existence of every kind of organism. Plants and animals have unique and diverse life cycles. (3-LS1-1)

EGG An adult female ladybug lays its eggs on a leaf.

LARVA The ladybug **larva** may eat small insects. It sheds its outer covering as it grows.

ADULT An adult ladybug has wings and can fly. It looks very different from the other stages.

PUPA The ladybug changes form during the **pupa** stage.

Wrap It Up!

1. **List** What are the stages of a ladybug's life cycle?

2. **Contrast** List some differences between the pupa and the adult stages in the ladybug life cycle.

Life Cycle of a Leopard Frog

Like ladybugs, frogs go through a life cycle in which the animals look different at different stages. Look at the photos of the life cycle of the leopard frog. Trace the diagram of its life cycle with your finger as you read about each stage.

Leopard frogs use their strong hind legs to escape danger.

DCI LS1.B: Growth and Development of Organisms. Reproduction is essential to the continued existence of every kind of organism. Plants and animals have unique and diverse life cycles. (3-LS1-1)

EGG An adult female leopard frog lays its eggs in a pond or swamp.

TADPOLE A **tadpole** has a tail and no legs. It lives underwater and breathes through gills.

ADULT An adult leopard frog breathes air and is often found on land or swimming in water.

YOUNG FROG The young leopard frog begins to grow legs. Its tail begins to shorten.

Wrap It Up!

1. **Recall** What are the stages in the life cycle of a frog?

2. **Contrast** Describe some differences between the tadpole stage and the adult stage of the frog.

Life Cycles

? **How can you show the sequence of stages in the life cycle of a spotted salamander?**

Spotted salamanders are close relatives of frogs. Like frogs, spotted salamanders live in two different habitats during their life cycles. Adult salamanders live on land, but young salamanders live in water. In this exercise, you will make a model of the life cycle of a spotted salamander.

Materials

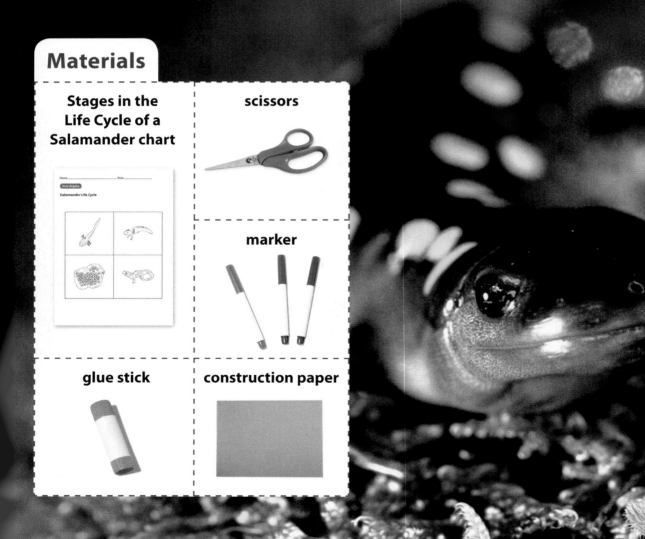

Stages in the Life Cycle of a Salamander chart

scissors

marker

glue stick

construction paper

DCI LS1.B: Growth and Development of Organisms. Reproduction is essential to the continued existence of every kind of organism. Plants and animals have unique and diverse life cycles. (3-LS1-1)

1 Look at the chart of the stages in the life cycle of a salamander. Use the Internet or other resources to find out about when each life stage begins and how long it lasts. Record the information in your science notebook.

2 Cut out the different stages of the salamander life cycle.

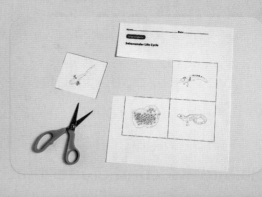

3 Place the steps on the construction paper in order. Check to be sure they are in the correct position. Then glue the steps to the construction paper.

4 Use the marker to draw arrows between the steps. Add a title to the life cycle diagram.

Wrap It Up!

1. **Describe** Where does the larva of a spotted salamander live?

2. **Compare** How are the life stages of the spotted salamander similar to the life stages of a frog?

3. **Contrast** How are the life stages of a spotted salamander different from those of a frog?

Develop a Model

You have read about different examples of life cycles. Now it's your turn to investigate, describe, and contrast two of them. Imagine that you have been asked to explain to a second grader that different living things can have very different life cycles. You choose two from the following examples:

- a gray whale
- a toad
- a monarch butterfly
- a tomato
- a dandelion

To help you describe the phenomena, you decide to develop models of them.

1. **Research the models.**
   Choose two of the plants or animals listed above. Use library books or the Internet to find out about the life cycles of the organisms you have chosen. Write down what you learn.

2. **Construct explanatory models.**
   Design models that use your research. Draw or describe your models. Explain all the life cycle stages you are going to include for both organisms.

3. **Construct your models.**
   Make your models. You may make more detailed drawings, or make physical models out of clay or other materials that show the patterns of change.

PE 3-LS1-1. Develop models to describe that organisms have unique and diverse life cycles but all have in common birth, growth, reproduction, and death.
NS Science Knowledge Is Based on Empirical Evidence. Science findings are based on recognizing patterns. (3-LS1-1)

4. **Analyze and revise your models.**

   Use the Internet or other resources to find out more about the life cycles of each organism. You might research about when each stage of the life cycle begins, about how long each stage lasts, or if the diet of an animal changes as it grows through its life cycle. Analyze your models. Add new information to make your models more complete.

5. **Share and explain your models.**

   Think about what you know about life cycles. Make predictions about what your classmates' life cycles will look like. Then share your models. Explain each stage. Compare and contrast your models. How are the life cycles of your two organisms different? How are they alike? What stages do they have in common? Expand your comparison to include the models made by your classmates.

Inherited Traits: Looks

Did you know that not all potatoes are brown? Potatoes can be sorted by color. Look at the large photograph. These potatoes have red, orange, white, yellow, and purple skins! The color is a **trait,** or characteristic, of the potatoes.

All living things have traits. Where did these potato traits come from? The traits of color and shape came from the parent plants. Traits that are passed down from parents to offspring are called **inherited traits.**

Corn can grow in many different colors on the same cob.

Tomatoes grow in many varieties, too. Most of their looks are inherited traits.

PE 3-LS3-1. Analyze and interpret data to provide evidence that plants and animals have traits inherited from parents and that variation of these traits exists in a group of similar organisms.

Analyze and Interpret Data

When you squeeze a snapdragon flower from the sides, it looks like a dragon that opens and closes its mouth. Snapdragons can vary in other traits, such as color. Observe the characteristics of the parent snapdragons and their offspring below.

PARENTS

OFFSPRING

1. What evidence can you provide to show that the offspring have inherited traits from their parents?

2. What evidence can you provide to show that the offspring have traits that vary from each other?

A potato's inherited traits include its size, shape, and color. The color can vary on the inside as well as the outside.

Wrap It Up! My Science Notebook

1. **Define** What is an inherited trait?

2. **List** List some inherited traits of potatoes.

3. **Generalize** List two other traits of tomatoes and corn besides color.

For the Love of Bugs

Some entomologists study insect behavior. Others look for patterns in tiny physical traits, such as wing or antennae shape, to help them identify and classify the insects.

DCI LS3.B: Variation of Traits. Different organisms vary in how they look and function because they have different inherited information. (3-LS3-1)
CCC Patterns. Similarities and differences in patterns can be used to sort and classify natural phenomena. (3-LS3-1)
NS Scientific Knowledge is Based on Empirical Evidence. Science findings are based on recognizing patterns. (3-LS1-1)

As a child, Jeurel watched ants crawl through the holes in the floor of her house. She knew the ants were communicating with each other.

Jeurel Singleton grew up in Arkansas. In 1965, she boarded a bus. She was going to college in North Dakota. She wanted to become an insect scientist, or entomologist. During the three-day trip, she did not eat or drink. Jeurel grew up during segregation. Many states had laws that kept people apart based on race. Black people had to sit at the back of a bus, while white people sat in the front. Black people could not use the same water fountains or bathrooms as white people. Signs marked what black people could use. "If you didn't see a sign," she said, "you didn't get off the bus."

When she first arrived in North Dakota, Jeurel had no place to stay. She camped in a cemetery for two weeks. She ate weeds and crickets, which she boiled in a tin can. When she finally moved into student housing, she was hungry. But she was excited that she would soon study insects.

At college, Jeurel became the entomologist she always wanted to be. She is now retired, but volunteers at zoos and schools. She still shares her collections and passion for bugs.

Wrap It Up!

1. **Summarize** What hardships did Jeurel face as she became an insect scientist?

2. **Observe** How do the butterflies in the collection vary in looks?

3. **Apply** What traits might scientists use to sort the insects?

Inherited Traits: Functions

Color in potatoes is an inherited trait that mainly affects how the potatoes look. But traits can also serve functions. For example, the shape of a bird's beak is an inherited trait. The size and shape of a bird's beak help the bird catch and eat its food.

The birds shown on these pages eat different types of food. They need different types of beaks. Catching a fish in a marsh, like whooping cranes do, requires a long beak that can poke into shallow water. Eating meat and cracking seeds require different kinds of beaks.

The bald eagle inherits a hooked beak that helps the bird tear meat from its prey.

PE 3-LS3-1. Analyze and interpret data to provide evidence that plants and animals have traits inherited from parents and that variation of these traits exists in a group of similar organisms.

The tan whooping crane is a young bird. It has the same inherited beak shape as the white adult whooping cranes.

Analyze and Interpret Data

Parakeets have a variety of traits such as feather colors and markings. Their strong, curved beaks help them crack husks off of seeds. Observe the characteristics of the parent parakeets and their offspring in the photos below.

1. What evidence can you provide to show that the offspring have inherited traits from their parents?

2. What evidence can you provide to show that the offspring have traits that vary in function?

Wrap It Up!

1. **Recall** What makes beak shape an example of an inherited trait?

2. **Explain** How can traits be classified by their function?

Acquired Traits

Not all traits are inherited. Some traits are acquired.
Acquired traits are gained from the environment.
For example, animals can acquire traits from their
diet. A diet is all the foods an animal eats. Diet affects
an animal's body size, weight, and health.

In some animals, diet can even change body color.
Flamingos are born with white feathers. Flamingos
have diets of algae, insect larvae, and shrimp.

DCI LS3.A: Inheritance of Traits. Many characteristics of organisms are inherited from their
parents. (3-LS3-1) • Other characteristics result from individuals' interactions with the environment, which
can range from diet to learning. Many characteristics involve both inheritance and environment. (3-LS3-2)
DCI LS3.B: Variation of Traits. Different organisms vary in how they look and function because they have
different inherited information. (3-LS3-1) • The environment also affects the traits that an organism develops.
(3-LS3-2)

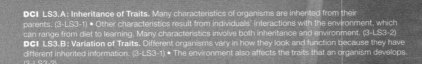

Some foods in their diet cause flamingos' feathers to turn pink. The more of these foods the birds eat, the pinker they get. Their color is an acquired trait.

These whiter flamingos have eaten less of the diet that causes the birds to change color.

These pinker flamingos have eaten more of the diet that causes the color change.

Wrap It Up! My Science Notebook

1. **Contrast** What is the difference between an inherited trait and an acquired trait?

2. **Identify** What is the evidence that pinkness in flamingos is an acquired trait?

3. **Explain** How might diet affect an animal's body weight?

More Acquired Traits

Traits can be acquired from other factors in the environment besides food. For example, weather can affect a plant's traits. A sunflower that gets a lot of sunlight will grow taller than a sunflower that only gets a little sunlight.

Some traits are inherited *and* affected by the environment. The general shape of a tree and its limbs are inherited traits. But a tree's branches can be bent and shaped by strong winds.

DCI LS3.A: Inheritance of Traits. Many characteristics of organisms are inherited from their parents. (3-LS3-1) • Other characteristics result from individuals' interactions with the environment, which can range from diet to learning. Many characteristics involve both inheritance and environment. (3-LS3-2)
DCI LS3.B: Variation of Traits. Different organisms vary in how they look and function because they have different inherited information. (3-LS3-1) • The environment also affects the traits that an organism develops. (3-LS3-2)

Strong winds blowing from one direction over time gave this tree its shape.

Wrap It Up! My Science Notebook

1. **Analyze** Which part of the shape of the tree shown on this page is inherited? Which part is acquired?

2. **Identify** What factor in the environment affects this tree's shape?

3. **Infer** Describe how you think this tree might look if it were growing in a place with little wind.

Learning

Many traits are physical characteristics, but other traits are actions or behaviors. Animals act to get food and meet their other needs. They interact with their environments. From those experiences, animals may change the way they behave. They can learn to behave differently.

Chimpanzees are skilled learners in the animal world. They not only learn how to find food, but they can learn how to use tools to do so! The ability to use a tool is not a skill a chimpanzee is born with. It is a behavior the chimpanzee acquires through learning. Animals can acquire many learned behaviors.

This chimpanzee uses a twig to collect termites from inside a termite mound.

This chimpanzee eats termites off of a twig that it used as a tool to collect them.

Wrap It Up! 📋 My Science Notebook

1. **Recall** What is behavior?

2. **Explain** What is something a chimpanzee can learn?

3. **Summarize** How can the environment affect the way an animal behaves?

93

STEM

RESEARCH PROJECT

Animal Behavior

When baby sea turtles hatch, they move in the direction of the brightest light. In their natural beach environment, they crawl toward the ocean, which reflects the light of the moon. This behavior is due to instinct, which is inherited behavior. Other animal behavior is learned. For example, young elephants learn social behavior that helps them get along with others in their group.

Some animal behavior is partly inherited and partly learned. Lion cubs have the instinct to hunt, but to become good hunters, they must learn skills by watching adult lions. Then they must practice.

The Challenge

Now you will tackle animal behavior on your own. You will choose an animal from a list. You must research and write about an interesting behavior of the animal. Your challenge is to describe whether the behavior is inherited, learned, or both.

PE 3-LS3-2. Use evidence to support the explanation that traits can be influenced by the environment.

African elephants learn how to live socially by watching older elephants. Without role models, young elephants often develop too much aggression.

STEM
RESEARCH PROJECT
(continued)

1 Select a topic. My Science Notebook

Choose one of the following animals:

- bearded capuchin monkey
- African elephant
- humpback whale
- orangutan
- common octopus

You will do research about the animal you choose, and you will describe how it behaves.

After you choose an animal, think about what you would like to know about it. Write questions to which you would like to have answers. One question to ask is: What interesting behavior does the animal have? Ask other *how, what, where, when,* or *why* questions. Identify key words in the questions. Use the key words to help guide your research.

2 Plan and conduct research.

Look in a library for nonfiction materials about the animal you chose. Use key words to search for resources. When you find a book that might be useful, use the parts of the book, such as the table of contents and chapter headings, to find information in the text.

In addition to the questions you wrote in Step 1, research the answers to the following questions:

- How does the behavior help the animal?
- Is the behavior learned, inherited, or both?
- What is one more interesting fact about the animal or the behavior?

Using your own words, take notes on information you find. Use graphic organizers to help you organize your notes.

A young orangutan stays near its mother for about the first five years of its life. This behavior is inherited.

3 Draft and finalize your report.

Your report will be in the form of a computer slide presentation. Using your research, describe the interesting behavior of the animal. What purpose does the behavior serve? Describe whether you think the behavior is learned or inherited, or if it is a mix of both. After you finish describing the behavior, add one more interesting fact about the animal or the behavior to your slide presentation.

Consider adding photos or videos of the animal to your slide presentation. At the end, make a list of the resources you used to find your information.

Review and revise the draft of your slide presentation to make it the best it can be. Make corrections as needed.

4 Present your report.

Practice giving your slide presentation aloud. You could practice in front of friends, or you could record yourself in a video. Ask for advice about how you could make your presentation better.

Share your slide presentation with the class. Put information in a logical order. Use descriptions, facts, and details to describe the animal and its behavior. Remember to say why the animal does the behavior and whether the behavior is inherited, learned, or both. Remember to also give an additional interesting fact about the animal or the behavior. Speak loudly so that everyone can hear you.

Listen as your classmates present their reports. How many different animal behaviors did your classmates identify and report on?

Environment and Traits

? **How does the amount of water a plant receives affect its growth?**

You have read that traits can be inherited, acquired, or both. Now it is your turn to test this idea. How much water a plant receives is a factor in the environment. In this investigation, you'll observe the effect of various amounts of water on plant growth.

Materials

wheatgrass seedlings	spray bottle with water	ruler	masking tape

PE 3-LS3-2. Use evidence to support the explanation that traits can be influenced by the environment.

1 Use masking tape to label one cup *water* and one cup *no water*. Observe the seedlings. Predict what will happen if one cup of seedlings gets watered and the other does not. Record your predictions.

2 Place your seedlings in a sunny spot. Measure and record the height of each seedling.

3 Spray the soil in the cup labeled *water* until the soil is slightly moist. Record the number of sprays you used.

4 Repeat step 3 every day for one week. Measure the height of the seedlings in both cups each day. Record your observations.

Wrap It Up! My Science Notebook

1. **Cause and Effect** How did the amount of water affect plant growth?

2. **Explain** Do your results provide evidence that traits can be affected by the environment? Explain.

3. **Conclude** Is seedling height an inherited trait, an acquired trait, or both? Explain.

Variation and Survival

An organism's traits can help it survive in its environment. Sea dragons have leaf-shaped structures all over their bodies. Each sea dragon looks a little different. The sea dragons that blend in best with seaweed are less likely to be seen and eaten. Sea dragons that stand out are in more danger of being eaten.
If certain traits are best for sea dragons, why don't they all have the same traits?

Sea dragons that blend in are more likely to survive, find mates, and reproduce.

DCI LS4.B: Natural Selection. Sometimes the differences in characteristics between individuals of the same

The answer is that environments change. The best traits in one place and time may not be best in another place and time. Variations help some of a type of living thing survive over time as things change. **Variations** are differences between individuals of the same type of organism.

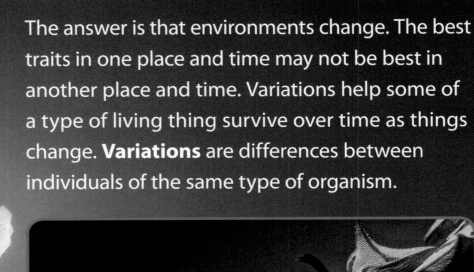

These thorn bugs all look very similar. The more each bug looks like a thorn, the less likely it is to get eaten by a bird.

Wrap It Up!

1. **Explain** Describe how traits of the sea dragon shown here help it survive.

2. **Infer** Suppose a few of the thorn bugs in the photo above were orange. What could you infer about the variation in thorn bug color?

Variation and Mates

Some traits help an organism find a mate. A male great frigatebird has a bright red pouch on his throat. He can fill the pouch with air so it swells up like a red balloon! This bright display gets the attention of females.

Female great frigatebirds do not have the showy patch of red. That helps the female and her young stay hidden from predators.

DCI LS4.B: Natural Selection. Sometimes the differences in characteristics between individuals of the same species provide advantages in surviving, finding mates, and reproducing. (3-LS4-2)

The male great frigatebird also uses behavior to attract a female. He opens his wings wide and shakes his head in a display like a dance. The female great frigatebird chooses the male with the most impressive display.

Wrap It Up! 📒 My Science Notebook

1. **Explain** What advantage does the trait of a bright red pouch give a male great frigatebird?

2. **Infer** What is the disadvantage of having the bright pouch?

Construct an Explanation

Katydids are related to grasshoppers and crickets. They live on shrubs and trees and feed on green leaves. In katydids, color is an inherited trait. The photographs on this page show actual colors of katydids. When katydids hatch from eggs they can be green, yellow, orange, or even pink! However, in adult katydids, the green form is far more common than other forms. Think of how variation can provide advantages in surviving, finding mates, and reproducing. Study the photos, then answer the questions.

Yellow adults were green as hatchlings.

Pink hatchlings remain pink in adulthood.

PE 3-LS4-2. Use evidence to construct an explanation for how the variations in characteristics among individuals of the same species may provide advantages in surviving, finding mates, and reproducing.

Wings of all colors of katydids have a veined pattern that resembles plant leaves.

Wrap It Up! My Science Notebook

1. **Construct an Explanation** Why might the pink form of katydid be less common in adults than in hatchlings?

2. **Cause and Effect** How does the trait of color help katydids survive?

Ecosystems

The forest is filled with living things—thousands of types of plants and animals. These living things are surrounded by the nonliving parts of the forest. Some nonliving parts are water, rocks, soil, and air.

The forest is one type of ecosystem. An **ecosystem** is a system that includes all the living and nonliving things in an area and the ways they interact. Populations, or groups of the same type of living thing, live in a wide variety of places. An ecosystem may be as large as a forest or the ocean. Other ecosystems, such as a pond or even a rotten log, are much smaller.

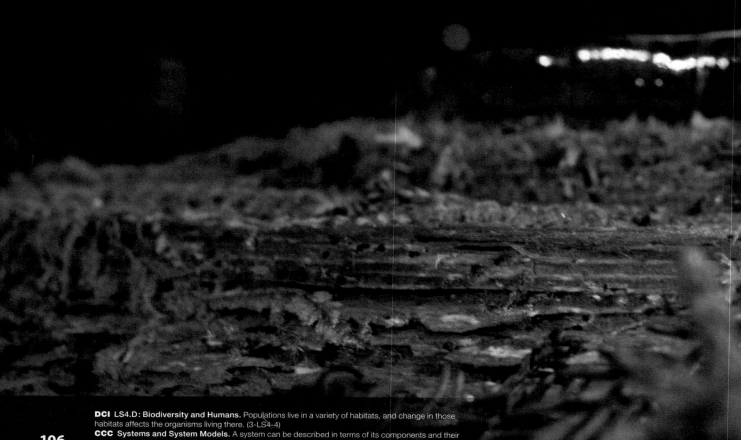

DCI LS4.D: **Biodiversity and Humans.** Populations live in a variety of habitats, and change in those habitats affects the organisms living there. (3-LS4-4)
CCC Systems and System Models. A system can be described in terms of its components and their interactions. (3-LS4-4)

The giant salamander is one living part of this forest ecosystem.

Wrap It Up! 📓 My Science Notebook

1. **Define** What is an ecosystem?

2. **Interpret** Look at the photo of the forest. What are some of the living things in this forest ecosystem?

3. **Infer** Name some of the nonliving things you cannot see that may be a part of this forest ecosystem.

BioBlitz

Field guides can help you identify wild plants and animals.

A hand lens can help you see the details needed for identification.

DCI LS4.C: Adaptation. For any particular environment, some kinds of organisms survive well, some survive less well, and some cannot survive at all. (3-LS4-3)
CCC Patterns. Similarities and differences in patterns can be used to sort and classify natural phenomena. (3-LS3-1)
CCC Systems and System Models. A system can be described in terms of its components and their interactions. (3-LS4-4)
NS Scientific Investigations Use a Variety of Methods. Science investigations use a variety of methods, tools, and techniques. (3-PS2-1)

What Is Citizen Science?

Scientists need your help. They want to know what lives in different ecosystems. They want to measure biodiversity. **Biodiversity** tells how many kinds of living things are found in a certain place. Biodiversity includes all of the plants and animals. It includes fungi, too.

Ordinary citizens can help measure biodiversity. In a BioBlitz, people count all of the things living in a certain area. Scientists use the findings in real scientific studies. Your class will conduct a BioBlitz to observe the biodiversity in your schoolyard, a park, or other area. You and your team will search for one group of living things. You might look for trees or wildflowers. Another team might look for birds, insects, or mushrooms. Look carefully. Try to identify each thing you find. Ask your teacher or another adult to take a picture. Write a description or draw a picture of everything you find. Be careful not to harm the living things. Leave the area as you found it.

Make a list of what you found. Use classroom resources or work with an adult to find the name of each living thing. Graph your class data. Draw a bar for each group of living things.

Wrap It Up!

1. **Describe** What was the hardest part of your BioBlitz? What was the most interesting thing you found?

2. **Analyze** Look at your class graph. Which group of organisms had the greatest biodiversity?

Forests Change

Fires, storms, and other natural events change the physical characteristics of a forest. A forest fire burns the trees and other plants. Afterwards, the ground is black and bare. There is less sheltered space. Such physical changes affect the organisms, or living things, that live there.

Soon after a fire, trees and wildflowers begin to grow.

DCI LS2.C: **Ecosystem Dynamics, Functioning, and Resilience.** When the environment changes in ways that affect a place's physical characteristics, temperature, or availability of resources, some organisms survive and reproduce, others move to new locations, yet others move into the transformed environment, and some die. (secondary to 3-LS4-4)

DCI LS4.D: **Biodiversity and Humans.** Populations live in a variety of habitats, and change in those habitats affects the organisms living there. (3-LS4-4)

CCC Cause and Effect. Cause and effect relationships are routinely identified and used to explain

There are no leaves for deer, rabbits, or insects to eat. Some of these animals move to other places to find food and shelter. Other animals die.

But a fire also helps some things grow. The burned wood adds nutrients to the soil. Grasses and wildflowers soon grow from the enriched soil. The heat of the fire makes the seeds of certain pine trees sprout. After a while, a new forest grows. Birds, deer, and other populations can move in.

A forest fire burns the trees and shrubs in the forest. There is little food for animals such as deer and rabbits.

Wrap It Up! My Science Notebook

1. **Describe** How does a fire change a forest?

2. **Cause and Effect** How does a fire affect the deer in a forest?

3. **Infer** After a forest fire, wildflowers soon begin to grow. How might the wildflowers eventually affect the number of rabbits in the area? Why?

111

Searching for Water

In the grasslands of East Africa, part of the year is rainy, and part of the year is dry. Water is a resource that plants and animals need. During the rainy season, water is plentiful. Water fills dried up rivers and ponds, and the soil is moist. Grasses and other plants grow well. Herds of animals such as zebras, wildebeests, and white-bearded gnu eat the grass.

During the dry season, the habitat changes. Rivers and pools dry up. The soil is so dry that grasses turn brown. The herds migrate to other places where there is grass to eat and water to drink. To **migrate** is to move to a different place to meet basic needs. Each year the zebras, wildebeests, and white-bearded gnu may travel up to a thousand miles to find water and green grass.

DCI LS2.C: Ecosystem Dynamics, Functioning, and Resilience. When the environment changes in ways that affect a place's physical characteristics, temperature, or availability of resources, some organisms survive and reproduce, others move to new locations, yet others move into the transformed environment, and some die. (secondary to 3-LS4-4)
DCI LS4.D: Biodiversity and Humans. Populations live in a variety of habitats, and change in those habitats affects the organisms living there. (3-LS4-4)
CCC Cause and Effect. Cause and effect relationships are routinely identified and used to explain change. (3-LS4-2), (3-LS4-3)

During the rainy season, watering holes provide animals with plenty of water to drink.

During the dry season, white-bearded gnu migrate in search of fresh grass and water.

Wrap It Up! My Science Notebook

1. **Identify** What are the two main seasons in the grasslands of East Africa?

2. **Cause and Effect** How does the dry season affect the grasses? Why?

3. **Infer** What might happen to the population of wildebeests if they did not migrate? Explain.

113

Changes in Temperature

In places with cold winter climates, winter air grows colder and the ground can freeze. There are fewer hours of daylight. With these changing physical characteristics, many populations of plants cannot grow. Some plants die in the winter but leave seeds that sprout into new plants in the spring.

DCI LS2.C: Ecosystem Dynamics, Functioning, and Resilience. When the environment changes in ways that affect a place's physical characteristics, temperature, or availability of resources, some organisms survive and reproduce, others move to new locations, yet others move into the transformed environment, and some die. (secondary to 3-LS4-4)

DCI LS4.D: Biodiversity and Humans. Populations live in a variety of habitats, and change in those habitats affects the organisms living there. (3-LS4-4)

Deciduous trees shed their leaves in the fall. They become dormant so they don't use as much energy. When it gets warmer, they grow new leaves.

Cold winter weather makes it hard for many animals to find food. Many birds migrate to places where more food is available. Other animals eat extra food in the fall and store it in their bodies as fat. Their bodies use the fat for energy during the cold winter.

During the summer, bees use nectar from flowers to make honey. The bees eat the honey through the winter.

The dormouse hibernates during the winter. While it **hibernates,** the animal's body does not use much energy, so it does not need to eat.

Wrap It Up!

1. **Describe** What are two ways that plants respond to changes in the environment, such as fewer hours of daylight and the cold weather of winter?

2. **Explain** How does hibernation help a dormouse survive?

3. **Generalize** How does the cold weather of winter affect the amount of food available to most animals?

Living Things Make Changes

Sometimes plants and animals change the environment. The pond in the picture was once a stream. Beavers used sticks and mud to build a dam. The dam holds back the water and turns the stream into a pond.

The new pond is a good habitat for beavers. A **habitat** is the place where a plant or animal lives and gets

Beavers use their teeth to chew down small trees into sticks to build their dams.

DCI LS2.C: Ecosystem Dynamics, Functioning, and Resilience. When the environment changes in ways that affect a place's physical characteristics, temperature, or availability of resources, some organisms survive and reproduce, others move to new locations, yet others move into the transformed environment, and some die. (secondary to 3-LS4-4)

DCI LS4.D: Biodiversity and Humans. Populations live in a variety of habitats, and change in those habitats affects the organisms living there. (3-LS4-4)

CCC Cause and Effect. Cause and effect relationships are routinely identified and used to explain change. (3-LS2-1), (3-LS4-2), (3-LS4-3)

everything it needs to survive. The deeper water of the pond lets beavers enter their dens underwater. This protects beavers from **predators,** or other animals that want to eat them. The beaver dam also affects other living things. The pond formed by the dam is a good habitat for marsh plants, turtles, dragonflies, and sunfish.

The beaver dam holds back the water and changes a stream into a pond. A small amount of water passes through the dam.

Wrap It Up!

1. **Describe** How does a beaver dam change a stream?

2. **Cause and Effect** How do beaver dams affect other animals in the ecosystem?

3. **Apply** Beavers use their teeth to cut down trees to make their dams. How might cutting down trees affect the animals in the nearby forest?

People Change Land

Human activities change the environment. People build roads and houses. They plant crops and build dams. These actions use space that once provided habitats for living things. When people cut down trees for wood, they change the forest. Some forest animals move to other places with trees, but others may die.

But people have ways to save wild plants and animals. People can plant young trees to replace the trees that are cut down for wood. Families can plant native trees and wildflowers in their yards. These plants provide food and shelter for birds, insects, and other animals.

When trees are cut down for wood, the forest environment changes.

DCI LS2.C: Ecosystem Dynamics, Functioning, and Resilience. When the environment changes in ways that affect a place's physical characteristics, temperature, or availability of resources, some organisms survive and reproduce, others move to new locations, yet others move into the transformed environment, and some die. (secondary to 3-LS4-4)
DCI LS4.D: Biodiversity and Humans. Populations live in a variety of habitats, and change in those habitats affects the organisms living there. (3-LS4-4)
CCC Cause and Effect. Cause and effect relationships are routinely identified and used to explain change. (3-LS4-2) (3-LS4-3)

Planting young trees helps a new forest grow.

Wrap It Up! 📓 My Science Notebook

1. **List** What are some human activities that change the environment?

2. **Cause and Effect** How does cutting down trees in a forest affect the animals that live there?

3. **Make Judgments** What do you think is a good way to protect the animals that live in forests? Explain.

SCIENCE
TECHNOLOGY
ENGINEERING
MATH

SPACE STATION PROJECT

Design a Seed Starter

Cutting down trees changes the environment and can harm other living things. Planting new trees and other plants is a solution to this problem. Plants provide many benefits. They provide oxygen that we breathe. They remove pollution from the air. They provide food and shade and create soil when they die. Plants create habitats for many living things.

Someday, people may create habitats for themselves in space. Plants will be a must! But how does being in space affect how seeds grow?

To find out, scientists are bringing seeds aboard the International Space Station. And what they are learning relates to growing plants here on Earth. By improving how seeds grow in space, we can improve how seeds grow on Earth, too.

PE 3–5-ETS1-2. Generate and compare multiple possible solutions to a problem based on how well each is likely to meet the criteria and constraints of the problem.
PE 3-LS4-4. Make a claim about the merit of a solution to a problem caused when the environment changes and the types of plants and animals that live there may change.

Two bags of tomato seeds were brought on board the International Space Station as a part of the Tomatosphere™ project.

The Challenge

Your challenge is to design and build a seed starter that makes it easy to sprout seeds indoors and transfer them outside while producing minimal waste. Your seed starter must:

- contain seedlings indoors near a window
- grow at least one 3-cm-tall seedling within two weeks
- produce no waste

Down to Earth Classes can register to plant seeds from the Tomatosphere™ project. The conditions the seeds are planted in will affect their chances of survival.

1 Define the problem. My Science Notebook

What problem can planting seeds help solve? State the problem clearly in your notebook. Criteria tell what your solution needs to do. Write your criteria.

Your teacher will show you materials to select among when designing your seed starter. You will also be given a spray bottle full of water and a space where your seed starters will be kept. Your teacher will tell you how much time you have to work on a design.

You cannot use any other materials, space, or time. These are the constraints of your solution. List the constraints.

2 Find a solution.

Look at the materials. How can you use them to make a seed starter? Think about these questions:

- What material will break down once it is planted in the soil?
- What shape will be best at holding the seed and its starting soil?
- Can any of the materials be changed or combined to improve the shape?

Be creative! Sketch your design, and present it to your team. Discuss each possible solution. Choose the one solution that will best meet the criteria. Draw your final design.

3 Test your solution.

Make a plan to test your solution. Your test will tell how well your seed starter meets the criteria. Include more than one seed starter in your test. Decide how you will water the starter soil and measure the seedlings once they sprout.

Construct your seed starters. Carry out your plan. Record all your observations in your notebook.

Discuss the results with your team. Talk about the other criteria. Did your design solve the problem?

4 Refine or change your solution.

Discuss ways you could improve your design. Draw your ideas.

Present your team's design to the class. Discuss its strengths and weaknesses. Tell about each criteria and whether your seed starter met it. Explain how you tested your solution. Answer questions about your design. Ask questions about the other teams' designs.

Make a plan to transfer the class's seedlings to an outdoor location and keep them alive. Your plants will make a positive change in the environment.

People Change Ecosystems

People change the physical characteristics of environments when they construct buildings and roads. These changes affect the populations of living things in the ecosystem.

This area was forest before these homes were built.

DCI LS2.C: Ecosystem Dynamics, Functioning, and Resilience. When the environment changes in ways that affect a place's physical characteristics, temperature, or availability of resources, some organisms survive and reproduce, others move to new locations, yet others move into the transformed environment, and some die. (secondary to 3-LS4-4)
DCI LS4.D: Biodiversity and Humans. Populations live in a variety of habitats, and change in those habitats affects the organisms living there. (3-LS4-4)

In cities, most of the ground is covered with buildings and pavement. Little can grow. Some people build gardens on flat rooftops. The plants give off oxygen and help clean the air. They help keep the buildings below them cooler. The plants and soil on the rooftop gardens capture rainwater that would otherwise wash away. These gardens can also provide people with fresh fruits and vegetables.

Paved streets and buildings absorb the sun's energy and make cities hotter than natural areas in the same region.

Plants on city rooftops provide a new habitat for birds and insects.

Wrap It Up!

1. **Describe** How do rooftop gardens change the rooftop environment?

2. **Compare** In summer, how would the temperature of a rooftop garden differ from that of a bare roof?

3. **Infer** How might building rooftop gardens affect the number of birds in a city? Why?

125

Compare Solutions and Make a Claim

1. Set the scene.

The Columbia River flows from the Canadian Rockies to the Pacific Ocean. Hundreds of streams and rivers flow into the Columbia making up a larger river system. Once the Columbia was filled with migrating salmon. Salmon spend much of their lives in the ocean. But they lay their eggs inland in freshwater streams. After young salmon hatch, they swim to the ocean. Years later, they return as adults to the freshwater streams to lay their eggs.

2. Define the problem.

In the past, more than ten million salmon migrated up the Columbia every year. Today, fewer than two million make the trip. Why? Engineers have designed huge dams on the Columbia and the rivers that flow into it. The dams produce electricity and store water for crops. But the dams keep the fish from swimming up and down the rivers. The drop in salmon population also affects salmon predators.

Many dams on the Columbia river and other rivers block migrating fish.

PE 3-LS4-4. Make a claim about the merit of a solution to a problem caused when the environment changes and the types of plants and animals that live there may change.
CETS Interdependence of Science, Engineering, and Technology. Knowledge of relevant scientific concepts and research findings is important in engineering. (3-LS4-4)

Salmon can jump up small waterfalls and swim against strong currents.

The John Day Dam is 56 meters (183 feet) tall. It completely blocks salmon migrating upstream. Water rushing through the power plant turbines also presents a danger to young salmon migrating downstream to the ocean.

Some adult salmon migrate for more than 900 miles! Many must pass through rapids to reach the streams where they lay their eggs.

3. Compare solutions.

Read the captions with these photos to find out what people are doing to try to increase the number of salmon in the Columbia River system. Some efforts help adult salmon migrate upstream to reproduce. Other efforts try to increase the numbers of young salmon and help them migrate to the ocean. Make a table that summarizes and compares the solutions.

4. Make a claim.

Suppose you are working with a team to design a new dam in the Columbia River system. You only have enough money to incorporate two of these four solutions into your design. Make a claim about which two solutions would be the best to include. Write a paragraph that explains why you chose to recommend those two solutions.

5. Support your claim.

Share your recommendation for the two best solutions with your classmates. Use details from your paragraph to defend why you believe your choices are the best for helping protect the salmon populations.

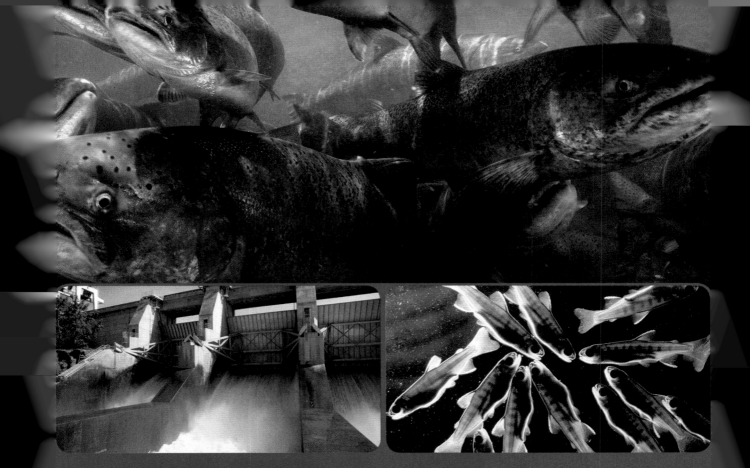

SPILLWAYS Spillways increase the amount of water that flows over the dams. They allow young fish to pass over the dams safely on their way to the ocean. The young fish also get a free ride in the fast currents.

FISH HATCHERIES Biologists raise salmon in hatcheries. The young salmon are released and swim downstream. This increases the number of salmon in the rivers.

FISH SCREENS Fish screens prevent fish from swimming into dangerous places. The screens shown here keep fish out of an unsafe channel. Similar screens are placed across turbine openings in dams. Young salmon caught in spinning turbines can be killed or stunned and then eaten by larger fish.

FISH LADDERS Many adult salmon are lost because they cannot swim up the water falling from the tall dams. A fish ladder is like a stairway with sets of small rapids. Salmon can jump from one step to the next. They can also swim through openings at the bottom of each step of the ladder.

Living in Groups

Some animals live alone. Others live in groups. Animal groups are called by different names—a flock of birds, a herd of elephants, a school of fish, a swarm of bees, a pod of whales. Not all animal groups are the same. For example, groups vary in size. A pack of wolves may have six to ten members, but some flocks of birds have more than one million members!

Meerkats may live in groups of 20 to 50 animals. They care for their young together and alert each other of danger.

DCI LS2.D: Social Interactions and Group Behavior. Being part of a group helps animals obtain food, defend themselves, and cope with changes. Groups may serve different functions and vary dramatically in size. (3-LS2-1)

Some animals live in groups because it helps them get food. Others live in groups to defend themselves or to help care for their young. Living in groups helps some animals cope with changes in the weather or their environment.

Emperor penguins live in Antarctica, where it is very cold. They flock together when it is time to produce their young.

Zebras are among many types of African animals that live in herds.

Wrap It Up! My Science Notebook

1. **Name** What are some of the names used for groups of animals?

2. **Contrast** Animal groups come in different sizes. Describe how the size of a wolf pack is different from the size of a large flock of birds.

3. **Summarize** How does living in groups help animals survive?

131

Getting Food

A bison runs through the snow, surrounded by a pack of wolves. A **pack** is a group of closely related animals that live and hunt together. Like a relay team, individual wolves have taken turns chasing the bison. Now the bison is too tired to run much longer. The bison tries to defend itself with its horns and hooves, but it cannot match the strength of the pack. Soon the pack closes in, and the hungry wolves share a meal of fresh meat.

Male cheetahs often hunt in small groups of two or three.

DCI LS2.D: Social Interactions and Group Behavior. Being part of a group helps animals obtain food, defend themselves, and cope with changes. Groups may serve different functions and vary dramatically in size. (3-LS2-1)

On its own, a wolf can catch only small **prey,** such as mice and rabbits. Working together, a pack of wolves can overtake large prey, such as moose, elk, and bison. These large animals may weigh ten times more than a single wolf. Wolves aren't the only animals that hunt in groups. Lions, hyenas, orcas, and army ants do, too. Being part of a group helps these animals get food.

When a bison becomes separated from its herd, it is in danger.

Wrap It Up! 📓 My Science Notebook

1. **Define** What is a pack?

2. **Contrast** What kind of prey can a single wolf catch? How is this different from the prey that a pack of wolves can catch?

3. **Generalize** How does hunting in groups help animals survive?

133

Protection and Defense

Living in groups helps some animals defend themselves against predators that would eat them. For example, fish such as the mackerel shown here swim in schools. A **school** is a group of fish that swim very close together. Swimming in a school provides "safety in numbers." There are many eyes to spot predators. The huge number of mackerel makes it hard for a predator to target a single fish.

Living in large groups can also protect animals that live on land. Wildebeests and zebras live in large herds. This helps protect them from lions and other predators. Flying in large flocks helps protect blackbirds and starlings from hawks.

DCI LS2.D: Social Interactions and Group Behavior. Being part of a group helps animals obtain food, defend themselves, and cope with changes. Groups may serve different functions and vary

Even though swimming in a school provides protection for most of the mackerel, dolphins will catch some of them!

Wrap It Up! My Science Notebook

1. **Define** What is a school of fish?

2. **Explain** How does swimming in a school help protect fish?

3. **Infer** How might swimming in a large school not help fish survive?

Coping with Change

When the conditions in a place change, groups of animals may travel together to a new place. Think about the changing seasons. In the fall, the weather in some areas gets cooler. Flocks of birds migrate to warmer places where they spend the winter.

Bees also move to cope with change. When a nest of bees gets too crowded, thousands of bees fly off in a swarm. A **swarm** is a large group of small animals moving together. The swarm includes a queen. When the bees find a good place, such as a hollow tree, they move in. Then the queen lays eggs and starts a new colony.

While this swarm of bees rests in a tree, scouts are searching nearby for a good place for their new hive.

DCI LS2.D: Social Interactions and Group Behavior. Being part of a group helps animals obtain food, defend themselves, and cope with changes. Groups may serve different functions and vary dramatically in size. (3-LS2-1)

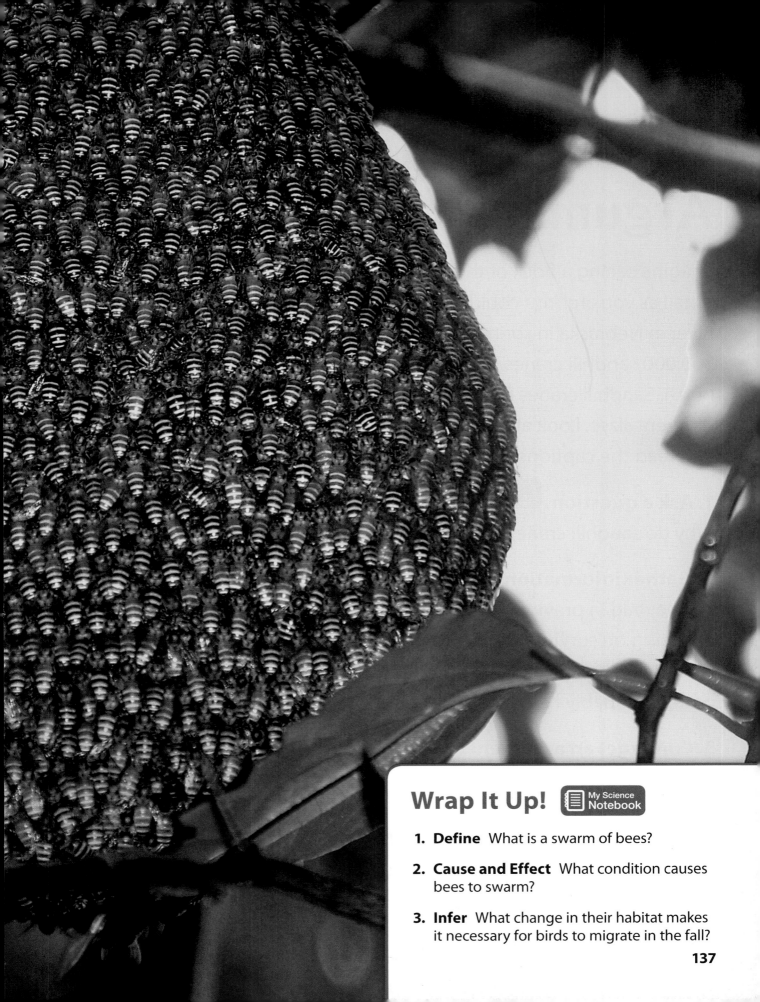

Wrap It Up! 🔲 My Science Notebook

1. **Define** What is a swarm of bees?

2. **Cause and Effect** What condition causes bees to swarm?

3. **Infer** What change in their habitat makes it necessary for birds to migrate in the fall?

Construct an  Argument

Imagine seeing a flock of thousands of birds almost as tall as you are! You could if you visited the Platte River in Nebraska in spring. That's when as many as 300,000 sandhill cranes come together as they migrate north. Sandhill cranes fly, flock, and nest in groups of different sizes. Look at the pictures of sandhill cranes and read the captions. Then, follow the steps below.

1. Ask a question.

Why do sandhill cranes form groups at various times?

2. Gather information.

You've read in previous lessons about the effects of grouping on survival. Recall that information as you read. Prepare a list of ways that migrating and living in groups probably helps sandhill cranes survive.

3. Construct an argument.

Write a paragraph that argues how the formation of groups helps sandhill cranes survive. Write about the different size groups they form at different times of the year, during and in between migrations.

PE 3-LS2-1. Construct an argument that some animals form groups that help members survive.

In summer, the sandhill cranes settle in marshes in the north to lay eggs and raise young. Pairs of parents care for the eggs and chicks. What are the advantages of both parents caring for the young?

In spring, sandhill cranes travel north to nest. In the fall, they travel south to warmer places to find food. The birds flock together in very large groups as they migrate. What are the advantages of resting in large flocks during migration?

In migratory flight, sandhill cranes fly in v-shaped formations. What are the advantages of flying in this type of group?

What are the strange creatures in this picture? They are animals that lived in the sea between 450 and 500 million years ago. Today many of these animals are extinct. When an animal is extinct, it is no longer living anywhere on Earth.

The hard shells of many trilobites were preserved in fossils. Can you find the large head and eyes on this trilobite fossil?

DCI LS4.A: Evidence of Common Ancestry and Diversity. Some kinds of plants and animals that once lived on Earth are no longer found anywhere. (3-LS4-1) • Fossils provide evidence about the types of organisms that lived long ago and also about the nature of their environments. (3-LS4-1)
CCC Scale, Proportion, and Quantity. Observable phenomena exist from very short to very long time periods. (3-LS4-1)

How do scientists observe phenomena from the ancient seas? They study fossils. **Fossils** are traces of plants, animals, and other organisms that lived long ago. Fossils provide evidence of ancient life.

Some fossils form when an animal dies and is buried in a layer of mud. Over very long time periods, the mud is pressed together and turns into rock. The rock preserves the shape of the animal.

Most fossils are the preserved remains of hard parts of an animal, such as its shell, bones, or teeth. Other fossils are preserved evidence, such as footprints or wormholes.

Five hundred million years ago, the sea was filled with animals without backbones. Some of these animals were ancestors of today's octopus and squid.

Wrap It Up! My Science Notebook

1. **Define** What is a fossil?

2. **Explain** How do fossils form?

3. **Infer** Dinosaur fossils usually show their bones but not their inner organs, such as the heart and lungs. Why do you think this is so?

Inside a Dinosaur Egg

Jack observed fossilized dinosaurs still in their eggs. This is one of his reconstructed fossils.

Unearthing a fossil dinosaur egg takes great care.

Jack is so famous that Dr. Grant in the first *Jurassic Park* movie is based on him.

DCI **LS4.A: Evidence of Common Ancestry and Diversity.** Fossils provide evidence about the types of organisms that lived long ago and also about the nature of their environments. (3-LS4-1)
CCC **Scale, Proportion, and Quantity.** Observable phenomena exist from very short to very long time periods. (3-LS4-1)
NS **Science Is a Human Endeavor.** Creativity and imagination are important to science. (3-LS4-1)

Life for Jack wasn't easy. In school, he wasn't good at reading, writing, and math. But Jack was good at science.

Jack Horner really digs dinosaurs. He found his first dinosaur fossil when he was 8 years old. He went on to make many dazzling dinosaur discoveries. As a scientist, Jack had the idea to crack open a dinosaur egg. He found that dinosaurs cared for their young. He showed that baby dinosaurs do not look like their parents. Jack became a famous paleontologist.

Life for Jack wasn't easy. In school, he wasn't good at reading, writing, and math. Classmates called him "dumb." But Jack was good at science and making science projects that won awards. Still, he got poor grades. Despite this, Jack went to college to learn about dinosaurs. He struggled there and dropped out. He still dreamed of making dinosaur discoveries.

Taking a job at Princeton University changed Jack's life. One day, he saw a poster. Its big letters said he could take a test to see if he had a learning disability. He took the test. It told him that he had dyslexia. People with dyslexia have trouble reading. This explained why Jack struggled in school.

Jack never let dyslexia stop him. He says that it helped him. He was creative and thought for himself. He says, ". . . I would challenge everything I was told, whether it was true or not."

Wrap It Up!

1. **Summarize** How did Jack turn a challenge into a success?

2. **Describe** What are some conclusions Jack reached about baby dinosaurs?

Fish in the Desert

Fossils provide evidence about plants and animals that lived long ago. They also provide evidence about the habitats in which they lived. Scientists study fossils to find out how the environment in a place has changed over time.

Today, many fish fossils can be found in a layer of rock called the Green River Formation. The formation spans through Utah, Colorado, and Wyoming. This area is now mostly a large, rocky, desert. But fish live only in wet places, such as lakes or the ocean. These fossils show that the whole area was once covered by water!

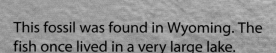

This fossil was found in Wyoming. The fish once lived in a very large lake.

DCI LS4.A: Evidence of Common Ancestry and Diversity. Some kinds of plants and animals that once lived on Earth are no longer found anywhere. (3-LS4-1) • Fossils provide evidence about the types of organisms that lived long ago and also about the nature of their environments. (3-LS4-1)
CCC Scale, Proportion, and Quantity. Observable phenomena exist from very short to very long time periods. (3-LS4-1)

This Utah canyon shows what much of the area looks like today.

Wrap It Up! 📋 My Science Notebook

1. **Contrast** How is the environment of Utah today different from the environment when the fossils in the picture formed?

2. **Draw Conclusions** Scientists have found fossils of clams in rocks at the top of mountains. What do these fossils suggest about the rocks?

Plants in the Antarctic

Sometimes the leaves and branches of plants are preserved in fossils. The fossil leaves shown here are from a type of extinct fern that grew as tall as trees! Like today's ferns, these plants grew in warm places.

This fern would have grown in a climate that was warm and humid.

DCI LS4.A: Evidence of Common Ancestry and Diversity. Some kinds of plants and animals that once lived on Earth are no longer found anywhere. (3-LS4-1) • Fossils provide evidence about the types of organisms that lived long ago and also about the nature of their environments. (3-LS4-1)
CCC Scale, Proportion, and Quantity. Observable phenomena exist from very short to very long time periods. (3-LS4-1)

Where was this fossil found? In Antarctica! Today Antarctica is so cold that the land is frozen all year long. Huge layers of ice cover most of the continent.

The fern fossil provides evidence that the rocks in Antarctica came from a place with a much warmer climate. Scientists can observe that Antarctica was once warm enough for ferns and forests to grow.

Antarctica is now far too cold for most plants to survive.

Wrap It Up! 🗒 My Science Notebook

1. **Identify** Where do ferns grow today?

2. **Describe** Contrast the present environment of Antarctica with the environments where ferns grow today.

3. **Infer** Why do you think fossil ferns can be found in a place that is frozen all year long?

147

Investigate

Fossils

? **How can you use a model to infer about past environments?**

Layers of rock often contain fossils. These fossils give scientists clues to the past. You will make a model of rock layers that contain fossils. First read the Fossil Environment chart below.

FOSSIL ENVIRONMENT

Layer Color	Environment When Layer Formed
Green	Dry land; warm temperatures
Yellow	Ice-covered land; dry, cold climate
Red	Warm, shallow, ocean water; warm, humid climate
Tan	Cool, freshwater lake; warm temperatures

Materials

4 lumps of clay	4 small objects

plastic knife	craft stick	tooth pick

DCI LS4.A: Evidence of Common Ancestry and Diversity. Fossils provide evidence about the types of organisms that lived long ago and also about the nature of their environments. (3-LS4-1)
CCC Scale, Proportion, and Quantity. Observable phenomena exist from very short to very long time periods. (3-LS4-1)

1 Flatten one lump of clay. This represents a layer of soil laid down millions of years ago. Press an object into the clay. The object represents an animal that died. Over time the soil formed rock. The animal became a fossil.

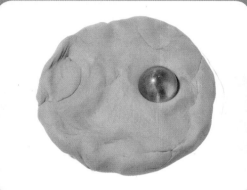

2 Add another layer of clay. This represents soil that covered the first layer and became rock. Press an object into this layer. Add two more layers with objects.

3 Exchange model rocks with another group. Cut through all of the layers. Draw the layers of the model in your notebook.

4 Use a toothpick or craft stick to remove the model fossils. Draw each one in the layer in which it was found. Use the chart to learn about each organism's environment. Record the data on your drawing.

The hard shells of sea animals can form very clear fossils.

Wrap It Up!

1. **Conclude** Describe the environment of animal fossils found in the red layer of rock.

2. **Infer** What can you infer about how the environment of the area represented by these layers changed over the years?

Analyze and Interpret Data

Environments on Earth change over long time periods. But some patterns in nature are consistent. Fish survive in water. Dragonflies survive on land. Their fossils tell a story.

Fossils are found in many parts of the world. The pictures show fossils that have been found in some of the places on the map. Look at the fossil pictures and read the captions. Using places on the map, find evidence about how the environment is now different from long ago.

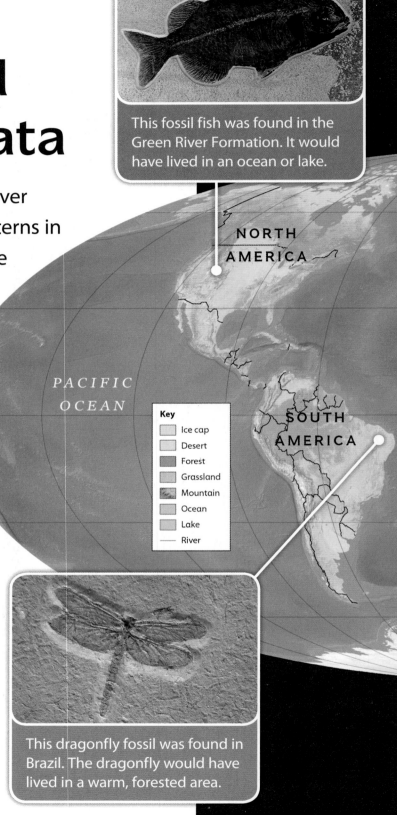

This fossil fish was found in the Green River Formation. It would have lived in an ocean or lake.

Key
- Ice cap
- Desert
- Forest
- Grassland
- Mountain
- Ocean
- Lake
- — River

NORTH AMERICA

PACIFIC OCEAN

SOUTH AMERICA

This dragonfly fossil was found in Brazil. The dragonfly would have lived in a warm, forested area.

PE 3-LS4-1. Analyze and interpret data from fossils to provide evidence of the organisms and the environments in which they lived long ago.
NS Science Knowledge Assumes an Order and Consistency in Natural Systems. Science assumes consistent patterns in natural systems. (3-LS4-1)

150

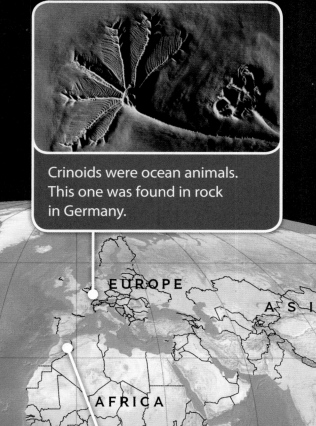

Crinoids were ocean animals. This one was found in rock in Germany.

This fossil arthropod was found in China. The segmented animal would have lived at the sea floor.

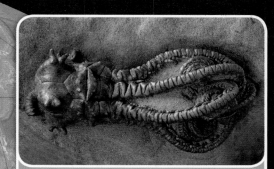

This crinoid was found in rock in Australia.

EUROPE

ASIA

AFRICA

ATLANTIC OCEAN

INDIAN OCEAN

AUSTRALIA

N
W — E
S

| 0 | 1,000 | 2,000 Miles |

| 0 | 1,000 | 2,000 Kilometers |

ANTARCTICA

This trilobite, an ocean animal, was found in Morocco.

Wrap It Up! 📖 My Science Notebook

1. **Interpret Maps** What do the different colored regions on the map represent?

2. **Interpret Data** Which of the fossils shown here came from a land area that was once covered in water? How do you know?

3. **Analyze** What kind of fossil was found in South America? What does this fossil show about the environment of that area a long time ago?

Cold or Warm?

Plants and animals need certain things from their environment to survive. Living things live where their needs can be met.

Polar bears live in the far north where the weather is cold. Their thick fur and body fat keep them warm. Polar bears spend most of the year on the ice that floats on the Arctic Ocean. Sea ice is the best place to catch the seals that swim in the cold water.

Gila monsters live where the weather is usually warm. Their bodies do not produce much heat. Instead, their body temperature depends on the temperature of their surroundings. If it is too cold, a lizard such as the gila monster cannot move fast enough to catch insects or avoid predators.

The polar bear is well suited for the cold climate, both on land and in icy water.

DCI LS4.C: Adaptation. For any particular environment, some kinds of organisms survive well, some survive less well, and some cannot survive at all. (3-LS4-3)

Gila monsters warm themselves in sunlight. They catch insects that live around the desert plants.

Wrap It Up! 📓 My Science Notebook

1. **Describe** Where do polar bears live? What is the temperature like there?

2. **Explain** How are polar bears able to survive in their environment?

3. **Draw Conclusions** Could a lizard survive where polar bears live? Why or why not?

153

Wet or Dry?

Leopard frogs live where it is wet. You might see one jumping at the edge of a pond or in a moist meadow. Like most frogs, they lay their eggs in water. The young frogs, or tadpoles, live underwater and breathe with gills. Adult leopard frogs have lungs and can breathe air. Gases can also move through the skin.

The leopard frog's smooth skin would dry out quickly in a desert environment.

DCI LS4.C: Adaptation. For any particular environment, some kinds of organisms survive well, some survive less well, and some cannot survive at all. (3-LS4-3)

For this to happen, they need to keep their skin moist. If a leopard frog's skin gets too dry, the frog will die.

Unlike leopard frogs, camels are well suited for living in dry places. Camels can survive for weeks without drinking water. When they find water, camels can drink many gallons at a time. Their bodies store it for later. Their thick, tough lips let them eat thorny desert plants.

Camels can survive many weeks without food. They use fat stored in the humps on their backs for energy.

Wrap It Up!

1. **Recall** Where do most frogs lay their eggs?

2. **Explain** How do camels survive in deserts?

3. **Infer** Could a leopard frog survive where a camel lives? Why or why not?

155

Light or Dark?

Just like temperature and moisture, different living things need different amounts of light to survive. Plants use sunlight to make food, so plants cannot live in totally dark places. Animals such as hawks, hummingbirds, and butterflies need bright light to find food. Bats, owls, and moths can find food without much light.

The amount of light varies in the ocean, too. The surface of the ocean is brightly lit. Seaweed and corals need bright light to survive. They live near the surface.

The black-eyed squid has large eyes compared to the rest of its small body. It can see where there is little light.

 DCI LS4.C: Adaptation. For any particular environment, some kinds of organisms survive well, some survive less well, and some cannot survive at all. (3-LS4-3)

But sunlight cannot reach below about 200 meters (656 feet). It is always dark there. Even so, many animals live in the dark parts of the ocean. Some of these creatures, such as the deep sea angler, make their own light! The fish has an organ that glows in the dark. It uses this organ to lure shrimp and fish into its mouth.

The verbena plant needs light to make its own food. The blue morpho butterfly needs light to find the flowers that provide it with nectar.

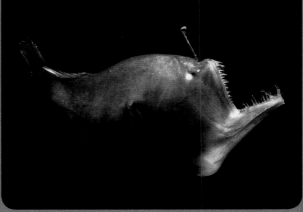

Without the photographer's light, only the lure dangling atop the deep sea angler's head would be visible.

Wrap It Up! My Science Notebook

1. **Recall** What are some animals that need bright sunlight to find food?

2. **Explain** Why do plants need sunlight?

3. **Evaluate** Could plants grow in the deep ocean? Explain.

Construct an Argument

Animals live in many different habitats. Some animals live where it is wet; others live only where it is dry. Some live where it is cold and dark; others live where it is warm and sunny. Look at the pictures of animals shown here and read the captions. Then follow the steps below.

1. **Ask a question.** 🔲 **My Science Notebook**
   In what kind of habitat do the particular animals shown here survive well?

2. **Read and observe.**
   You've read in previous lessons about animals that survive well in very different habitats. Recall that information as you read these pages. Prepare a list of the kinds of things each animal needs to survive well.

3. **Construct an argument.**
   Use your list as evidence. Construct an argument that describes the characteristics of the habitat in which each animal can survive well. Write a paragraph that tells how the animal's needs are met. Then tell how a change in the habitat would affect the animal's survival.

Mallard ducks have webbed feet and wide yellow bills. They eat the roots and stems of plants that grow underwater. They also eat small animals that live underwater.

PE 3-LS4-3. Construct an argument with evidence that in a particular habitat some organisms can survive well, some survive less well, and some cannot survive at all.

This strange creature is a star-nosed mole. The mole is practically blind. But its highly sensitive nose has many long feelers that help it find worms to eat. How do you think it uses its big, clawed feet?

A muskox is a large animal with long, shaggy hair. It eats grasses, mosses, and lichens. Its sharp hooves help it dig through snow to find these plants.

Marine Ecologist

Enric Sala grew up on the coast of Spain. He loved the sea. But he also saw how people were changing the characteristics of the sea. They were polluting the water. They were taking too many fish. Enric decided to spend his life working to save the health of the ocean.

Today Enric is a marine ecologist. He studies marine ecosystems—the communities of living things in the ocean. Enric leads scientific explorations to some of the most unspoiled parts of the ocean. He and his team have discovered crystal clear water, a coral reef growing deeper in the ocean than any other, new kinds of fish, and an amazing number of sharks!

Enric takes photos of organisms in the ocean ecosystems he studies.

Enric shares what he discovers in National Geographic publications and television programs. He says, "I want to show the world what the ocean was like hundreds of years ago and why we have to preserve it."

His work has inspired the leaders of some countries to set aside marine protected areas. He hopes that his work will help save some of the last untouched marine ecosystems on Earth.

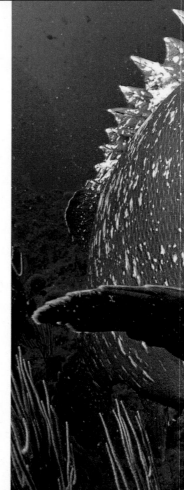

DCI LS2.C: Ecosystem Dynamics, Functioning, and Resilience. When the environment changes in ways that affect a place's physical characteristics, temperature, or availability of resources, some organisms survive and reproduce, others move to new locations, yet others move into the transformed environment, and some die. (secondary to 3-LS4-4)

NS Scientific Knowledge Assumes an Order and Consistency in Natural Systems. Science assumes consistent patterns in natural systems. (3-LS4-1)

Enric Sala is a marine ecologist and National Geographic Explorer. He works with organizations and governments to protect ocean ecosystems.

The endangered dusky grouper is one of the animals that Enric's work might help to save.

Check In  My Science Notebook

Congratulations! You have completed *Life Science*. Let's reflect on what you have learned. Here is a checklist to help you judge your progress. Look through your science notebook to find examples of each item in the list.

What could you do better? Write it on a separate page in your science notebook.

▼ Read each item in this list. Ask yourself if you think you did a good job of it.

For each item, select the choice that is true for you: A. Yes B. Not Yet

- I defined and drew pictures of science vocabulary, science concepts, and main ideas.
- I labeled drawings. I included captions and notes to explain ideas.
- I collected objects, such as photos and magazine or newspaper clippings.
- I used tables, charts, or graphs to record observations and data in investigations.
- I recorded evidence for explanations and conclusions in investigations.
- I described how scientists and engineers answer questions and solve problems.
- I asked new questions.
- I did something else. (Tell about it.)

Reflect on Your Learning My Science Notebook

1. Choose one investigation that you found most interesting. Explain what made it most interesting to you.

2. How has what you learned changed the way you think about the world around you?

Andrés records observations from the rain forest in Peru to bring back to his lab.

Andrés Ruzo Geoscientist
National Geographic Explorer

Let's Explore!

In *Nature of Science,* you learned that observations are important in science. Scientists often use tools to collect data. I collect data using my thermal camera, a special tool that lets me see the heat coming from Earth. I also record my observations as numbers whenever I can. This helps me analyze the data, or look for patterns. From patterns, we can make explanations and predictions. We can solve problems. Look for tools that scientists use to collect data in the lessons ahead.

Earth science is the study of Earth and its atmosphere. Here are some questions you might answer in *Earth Science*:

- What does a barometer measure?

- When air pressure drops, what kind of weather might you expect?

- How is the climate in Giza, Egypt, different from the climate in Minneapolis, Minnesota?

- How do lightning rods protect buildings?

- How can you design a house to protect it from hurricane damage?

Look at the notebook examples for some ideas. Let's check in again later to review what you have learned!

Weather Facts

1. Air pressure is the force of air pushing on Earth. When air pressure rises, the weather will be fair. I can use a barometer to measure air pressure.

2. A weather map shows temperature, precipitation, and warm and cold fronts. I can look at weather maps to see what kind of weather is forecasted for my area.

My New Questions

1. How accurate are the weather forecasts in my area?

2. How can I use weather instruments to predict what the weather might be like tomorrow?

3. What kind of plants are best suited for the climate in my area?

4. What are ways to stay safe from any weather hazards that are common in my area?

▶ Use your notebook to reflect on what you've done and learned.

My Weather Report

Day	Symbol	Weather
1		Sunny
2		Partly cloudy
3		Partly cloudy
4		Rainy
5		Thunderstorms

Earth Science

Weather and Climate

This farm sits in the path of an oncoming storm.

Weather

How would you describe the weather? **Weather** describes what the conditions in the air outside are like at a certain time and place. You might say it was sunny or cloudy, hot or cold, windy or calm, or dry or wet. You would probably use several of those words.

How would you describe the weather in this photograph? Do you think the temperature is more likely to be warm or very cold?

How would you describe the weather in this photograph? How is it different from the weather in the photograph above?

DCI ESS2.D: Weather and Climate. Scientists record patterns of the weather across different times and areas so that they can make predictions about what kind of weather might happen next. (3-ESS2-1)
NS Science Is a Human Endeavor. Science affects everyday life. (3-ESS3-1)

weather changes from day to day, even hour to hour. Weather also changes with the seasons. Scientists measure these changes and record these patterns so that they can predict what the weather might be like in the future.

Stormy weather can be dangerous. During thunderstorms, stay indoors and away from windows!

Wrap It Up! My Science Notebook

1. **Define** What is weather?

2. **Explain** Tell three ways in which weather can change.

3. **Describe** What is the weather like in the large photo on these two pages?

Weather Measurements

Scientists use different instruments to measure changes in weather. They record patterns of weather across different times and areas. They use this information to predict what kind of weather might happen next.

THERMOMETER

A **thermometer** measures air temperature.

WIND VANE

A **wind vane** shows the direction from which the wind is blowing.

BAROMETER

A **barometer** measures air pressure.

RAIN GAUGE

A **rain gauge** measures rainfall.

DCI ESS2.D: **Weather and Climate.** Scientists record patterns of the weather across different times and areas so that they can make predictions about what kind of weather might happen next. (3-ESS2-1)

One measurement is temperature. Scientists measure temperature to find out how hot or cold the air is. Changes in temperature cause air to move.

Moving air is called **wind.** Wind can change speed and direction. The direction wind comes from can bring cooler air or rain with it. Scientists also measure how much **precipitation,** or water such as rain and snow, falls from clouds.

Air pressure is the force with which air pushes on Earth. When air pressure rises, weather will be fair. When it drops, weather will be cloudy and often stormy.

This scientist is using an instrument that measures both the speed and direction of the wind.

Wrap It Up! 🗒 My Science Notebook

1. **Explain** Tell why scientists record patterns of weather across different times and areas.

2. **Cause and Effect** How do changes in temperature affect air?

Investigate

Weather

? **How can you measure some changes in the weather?**

You have read about some of the instruments that scientists use to measure changes in weather and make predictions about what kind of weather might happen next. In this investigation, you'll make an **anemometer,** or tool that measures wind speed. You'll use your anemometer and a thermometer to record some changes in weather.

Materials

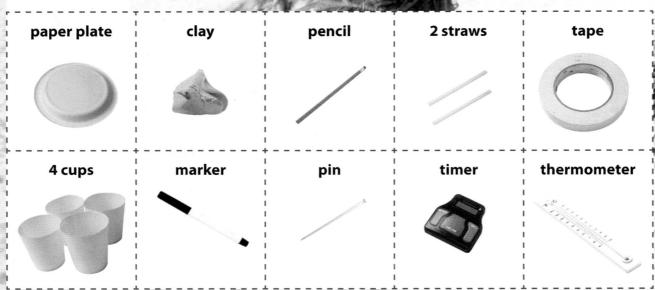

paper plate	clay	pencil	2 straws	tape
4 cups	**marker**	**pin**	**timer**	**thermometer**

DCI ESS2.D: Weather and Climate. Scientists record patterns of the weather across different times and areas so that they can make predictions about what kind of weather might happen next. (3-ESS2-1)

My Science Notebook

1 Follow your teacher's instructions for making the anemometer.

2 Place the anemometer in an open area outside. Each group should choose a different area. Use the timer to estimate wind speed by counting the number of times the cup with the X spins around in 1 minute. Record the data in your science notebook.

3 At the same spot, use the thermometer to measure air temperature. Record your measurement in degrees Celsius.

4 Repeat steps 2 and 3 at the same place and time each day for a week. Compare and contrast your data with those of other groups.

Wrap It Up! My Science Notebook

1. **Summarize** How did the data for wind speed and temperature change during the week?

2. **Explain** How did your weather tools help you to measure the weather conditions?

3. **Compare and Contrast** How were the data collected at different areas alike and different?

Patterns and Predictions

Scientists record data about weather conditions on maps. They use the maps to predict what the weather might be like in one hour, one day, or one week.

These weather maps show temperature and precipitation. They also show fronts. A **front** is a place where two very large masses of air meet. Weather events, such as the storm shown in the picture, can happen at a front.

Fronts can cause weather to change. A cold front brings cooler weather to an area. A warm front brings warmer weather.

These maps use symbols to show weather data. The keys explain what the symbols mean. How did weather conditions change from Day 1 to Day 2?

DCI ESS2.D: **Weather and Climate.** Scientists record patterns of the weather across different times and areas so that they can make predictions about what kind of weather might happen next. (3-ESS2-1)
CCC **Patterns.** Patterns of change can be used to make predictions. (3-ESS2-1), (3-ESS2-2)

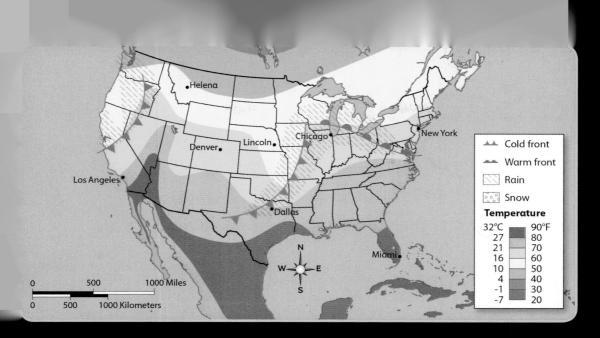

DAY 2

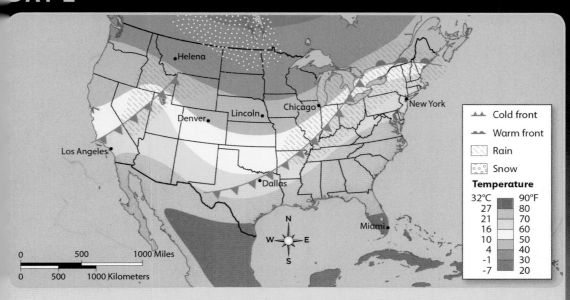

Scientists use maps of weather data to predict when storms may be coming.

Wrap It Up!

1. **Describe** In general, how did the fronts move from Day 1 to Day 2?

2. **Interpret Maps** Look at the Day 1 map. Describe the weather in Chicago on that day.

3. **Predict** Study both maps. Look for patterns. Tell what you think the weather will be like in Chicago on the day after Day 2.

The Pattern of the Seasons

In most places on Earth, weather changes with the seasons. In many areas, winter weather is cold and snowy. As winter changes to spring, the weather

SPRING In spring, the temperatures warm. Spring weather can be very windy and wet.

SUMMER Summer is the hottest season. In Wyoming, the summer weather is often dry.

These photos show Grand Teton National Park in Wyoming during different seasons.

DCI ESS2.D: Weather and Climate. Scientists record patterns of the weather across different times and areas so that they can make predictions about what kind of weather might happen next. (3-ESS2-1)
CCC Patterns. Patterns of change can be used to make predictions. (3-ESS2-1), (3-ESS2-2)

warms and can be very rainy. As spring changes to summer, the weather gets even warmer. Some places have very hot, dry summers. As summer changes to fall, the weather gets cooler and drier. As fall changes to winter, weather becomes colder again. This pattern repeats year after year.

FALL In fall, the weather cools. Fall weather in Wyoming is drier than spring and winter.

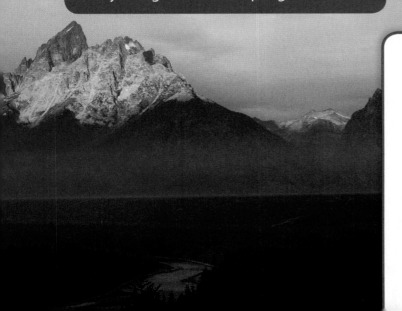

WINTER Winter is the coldest season. Snow falls in some areas in winter.

Wrap It Up!

1. **Sequence** Name the seasons in order, starting with winter.

2. **Contrast** Tell how weather differs in spring and summer.

3. **Predict** The average temperature of a city is 9°C (48°F) in winter and 29°C (83°F) in summer. Estimate what its average spring temperature might be.

Seasonal Changes

You might have a coat you wear only in winter because you know the weather is usually cold. But you know that on any given day it might also be sunny, rainy, or snowy.

SEASONAL PATTERNS IN CHARLOTTESVILLE, VIRGINIA.

	WINTER Dec. 22– Mar. 19	SPRING Mar. 20– June 20	SUMMER June 21– Sept. 21	FALL Sept. 22– Dec. 21
AVERAGE HIGH TEMPERATURE	9°C (48°F)	21°C (70°F)	28.5°C (83°F)	16°C (61°F)
AVERAGE LOW TEMPERATURE	-1°C (30°F)	9.5°C (49°F)	17°C (63°F)	5°C (41°F)
AVERAGE PRECIPITATION	20.68 cm (8.14 in.)	27.42 cm (10.80 in.)	33.78cm (13.80 in.)	26.24 cm (10.33 in.)

DCI ESS2.D: Weather and Climate. Scientists record patterns of the weather across different times and areas so that they can make predictions about what kind of weather might happen next. (3-ESS2-1)
CCC Patterns. Patterns of change can be used to make predictions. (3-ESS2-1), (3-ESS2-2)

Weather changes often, even within a season. So scientists collect a lot of data to describe the average weather for an area during a season. Look at what the weather data show about each season in Charlottesville, Virginia.

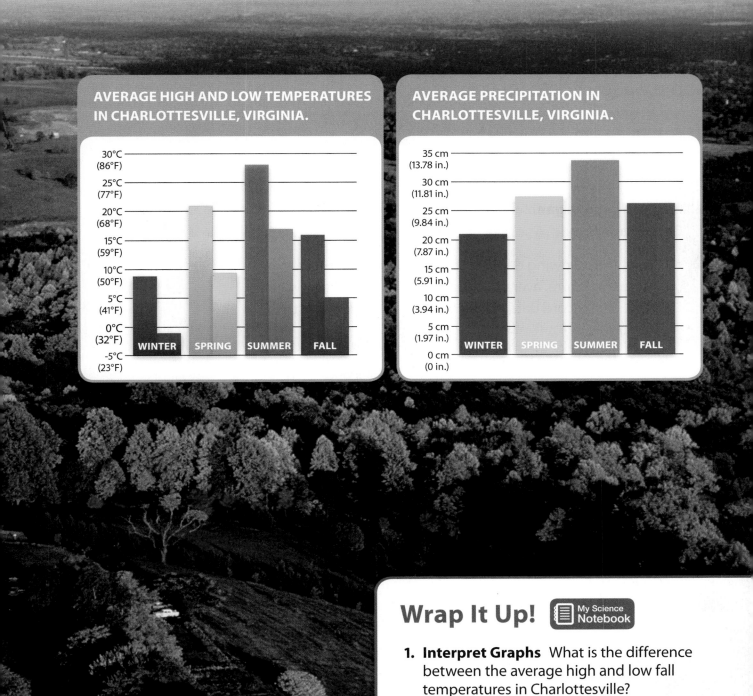

AVERAGE HIGH AND LOW TEMPERATURES IN CHARLOTTESVILLE, VIRGINIA.

30°C (86°F)
25°C (77°F)
20°C (68°F)
15°C (59°F)
10°C (50°F)
5°C (41°F)
0°C (32°F)
-5°C (23°F)

WINTER   SPRING   SUMMER   FALL

AVERAGE PRECIPITATION IN CHARLOTTESVILLE, VIRGINIA.

35 cm (13.78 in.)
30 cm (11.81 in.)
25 cm (9.84 in.)
20 cm (7.87 in.)
15 cm (5.91 in.)
10 cm (3.94 in.)
5 cm (1.97 in.)
0 cm (0 in.)

WINTER   SPRING   SUMMER   FALL

It is fall in the Blue Ridge Mountains near Charlottesville, Virginia.

Wrap It Up! 📓 My Science Notebook

1. **Interpret Graphs** What is the difference between the average high and low fall temperatures in Charlottesville?

2. **Predict** What might the average amount of precipitation be in Charlottesville next summer?

179

Represent Data

You have seen how graphs and tables were used to represent data. Now you will find and organize some weather data for your area.

1. **Ask a question.**
   How can you represent some weather data for one season in your area?

2. **Research and organize data.**
   Choose a season.

   - Collect the following data for that season: average low and high temperature, average precipitation, and average wind speed. Also record data on the direction the wind most often blows.

   - Organize your data in one or more tables.

   - Use colored pencils to make bar graphs of your temperature, precipitation, and wind speed data.

3. **Analyze and interpret data.**
   Meet with other students who researched data on the same season you chose. Compare your data. How is the data alike or different? Do more research until your group is satisfied that your data charts and tables are accurate.

4. **Present and explain.**
   As a group, present your season data to the class. Listen to other groups as they present their data. What is the typical weather like in each season in your area?

PE 3-ESS2-1. Represent data in tables and graphical displays to describe typical weather conditions expected during a particular season.

Explore on Your Own

How do weather conditions in your area compare to those in another area? Choose a location somewhere else on Earth. Predict the general weather conditions for each season. Then conduct research to find out how the typical weather conditions compare.

Temperature is weather data that people can easily collect at home.

Climate

Scientists summarize the yearly weather data in an area to describe that area's climate. **Climate** is the general pattern of weather in an area over a long period of time. Climate doesn't change very much from year to year.

The climate of an area can be very dry or very wet, or humid. Similarly, one climate might be very hot or very cold, while others have mild temperatures.

Portland, Oregon
AVERAGE JANUARY TEMPERATURE: **4.2°C** (39.6°F)
AVERAGE JULY TEMPERATURE: **20.1°C** (68.2°F)
AVERAGE ANNUAL PRECIPITATION: **92.1 cm** (36.3 in.)

Yuma, Arizona
AVERAGE JANUARY TEMPERATURE: **13.6°C** (56.5°F)
AVERAGE JULY TEMPERATURE: **34.2°C** (93.6°F)
AVERAGE ANNUAL PRECIPITATION: **7.7 cm** (3 in.)

DCI ESS2.D: **Weather and Climate.** Climate describes a range of an area's typical weather conditions and the extent to which those conditions vary over years. (3-ESS2-2)

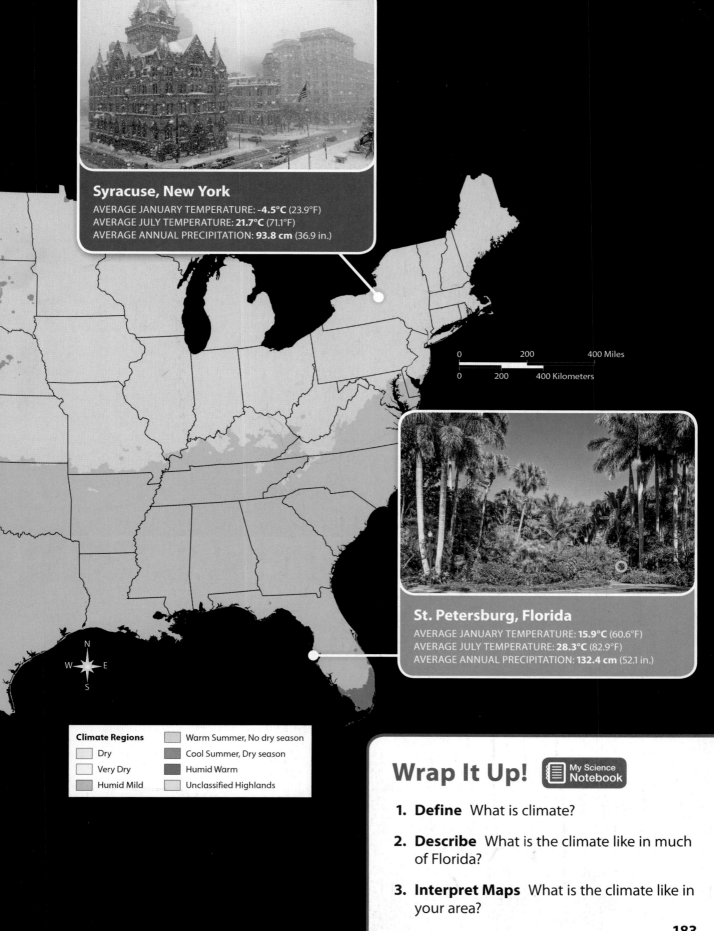

Syracuse, New York
AVERAGE JANUARY TEMPERATURE: **-4.5°C** (23.9°F)
AVERAGE JULY TEMPERATURE: **21.7°C** (71.1°F)
AVERAGE ANNUAL PRECIPITATION: **93.8 cm** (36.9 in.)

St. Petersburg, Florida
AVERAGE JANUARY TEMPERATURE: **15.9°C** (60.6°F)
AVERAGE JULY TEMPERATURE: **28.3°C** (82.9°F)
AVERAGE ANNUAL PRECIPITATION: **132.4 cm** (52.1 in.)

0   200   400 Miles
0   200   400 Kilometers

Climate Regions
- Dry
- Very Dry
- Humid Mild
- Warm Summer, No dry season
- Cool Summer, Dry season
- Humid Warm
- Unclassified Highlands

Wrap It Up! 📔 My Science Notebook

1. **Define** What is climate?

2. **Describe** What is the climate like in much of Florida?

3. **Interpret Maps** What is the climate like in your area?

183

Climate and Culture

Climate change affects the timing of floods, plant growth, and blooming.

The timing of bird, fish, and elk migration has changed in turn.

DCI ESS2.D: Weather and Climate. Climate describes a range of an area's typical weather conditions and the extent to which those conditions vary over years. (3-ESS2-2)
CCC Scale, Proportion, and Quantity. Observable phenomena exist from very short to very long time periods. (3-LS4-1)
NS Scientific Knowledge Is Based on Empirical Evidence. Science findings are based on recognizing patterns. (3-LS1-1) (3-PS2-2)

Native American people noticed the changing climate before scientists did.

Samantha Chisholm Hatfield grew up listening to old stories. She mostly ate clams, deer, elk, salmon, and smelt. As both a Cherokee and a member of the Siletz tribe, she grew up close to nature. Today, Samantha is a climate scientist. She uses her Native American heritage to help her as a scientist.

Climate is an area's weather over a long period of time. Weather can change daily or even hourly. Climate changes, too. But climate usually changes slowly. For decades, Earth's climate has been changing faster than it normally does. Native American people noticed the changing climate before scientists did. "Native people have a sort of alarm clock," Samantha says. They rely on nature's changing patterns. So they notice even small changes in those patterns.

One change the Siletz have seen is when some ants come out in the spring. They call them "eel ants." That's because the ants used to come out before the Siletz fished for eels. Now the temperatures start getting warmer earlier. The timing is off. This change affects when Native American people hunt, fish, forage, and plant crops. Samantha's knowledge of nature has helped her study climate change.

Wrap It Up!

1. **Summarize** How does Samantha's culture help her as a climate scientist?

2. **Explain** Describe one observation the Siletz people made, and explain how it indicated climate change.

Obtain and Combine Information

Imagine living in each city on the map. Use the map, photos and data to determine what the climate would be like in each city and how they compare.

Oulu, Finland
AVERAGE JANUARY TEMPERATURE: **−11.8°C** (10.8°F)
AVERAGE JULY TEMPERATURE: **15.7°C** (60.3°F)
AVERAGE ANNUAL PRECIPITATION: **45.3 cm** (17.8 in.)

Equator

0   1,000   2,000 Miles
0   1,000   2,000 Kilometers

Minneapolis, Minnesota
AVERAGE JANUARY TEMPERATURE: **−1.7°C** (28.9°F)
AVERAGE JULY TEMPERATURE: **26.9°C** (80.4°F)
AVERAGE ANNUAL PRECIPITATION: **68.4 cm** (26.9 in.)

Buenos Aires, Argentina
AVERAGE JANUARY TEMPERATURE: **23.5°C** (74.3°F)
AVERAGE JULY TEMPERATURE: **10.0°C** (50.0°F)
AVERAGE ANNUAL PRECIPITATION: **100.5 cm** (39.6 in.)

Giza, Egypt
AVERAGE JANUARY TEMPERATURE: **13.8°C** (56.8°F)
AVERAGE JULY TEMPERATURE: **27.9°C** (82.2°F)
AVERAGE ANNUAL PRECIPITATION **2.1 cm** (0.8 in.)

Magadan, Russia
AVERAGE JANUARY TEMPERATURE: **–17.2°C** (1.0°F)
AVERAGE JULY TEMPERATURE: **6.8°C** (44.2°F)
AVERAGE ANNUAL PRECIPITATION: **53.3 cm** (21.0 in.)

Alice Springs, Australia
AVERAGE JANUARY TEMPERATURE: **28.5°C** (83.3°F)
AVERAGE JULY TEMPERATURE: **11.5°C** (52.7°F)
AVERAGE ANNUAL PRECIPITATION: **28.1 cm** (11.1 in.)

Climate Regions

Dry	Warm Summer, No dry season
Very Dry	Cool Summer, Dry season
Humid Mild	Cold Polar–Tundra & ice
Humid Warm	Unclassified Highlands

Kinshasa, DR Congo
AVERAGE JANUARY TEMPERATURE: **25.2°C** (77.4°F)
AVERAGE JULY TEMPERATURE: **21.6°C** (70.9°F)
AVERAGE ANNUAL PRECIPITATION: **140.6 cm** (55.3 in.)

Wrap It Up!

1. **Interpret Maps** Use the map key to describe the climate of Minneapolis, Minnesota.

2. **Identify** Name two cities with similar climates. How do you know they have similar climates?

3. **Predict** What climate would you predict to find if you traveled toward the Equator?

187

Weather Hazards

Thunderstorms and hurricanes are types of weather that can damage property and harm people. Hurricanes are severe tropical storms that form over the ocean. They drop much rain in a very short time. When hurricanes reach the shore, they can cause water to rise into land areas that are normally dry. These floods can damage property. Floods are also dangerous for people who cannot get to higher ground.

Lightning can occur during storms over water and land.

DCI ESS3.B: Natural Hazards. A variety of natural hazards result from natural processes. Humans cannot eliminate natural hazards but can take steps to reduce their impacts. (3-ESS3-1)
NS Science Is a Human Endeavor. Science affects everyday life. (3-ESS3-1)

Hurricanes bring strong winds, too. Trees can bend and break, and buildings can be damaged by the force of a hurricane's winds.

The Jet Star roller coaster fell into the ocean during Hurricane Sandy. The pier it was sitting on collapsed.

Thunderstorms that are not as big as hurricanes can still cause damage. Rain from such storms can cause flooding. Lightning can also damage property.

Strong wind and heavy waves batter this pier during Hurricane Sandy.

Wrap It Up! 📓 My Science Notebook

1. **Identify** Name two types of hazardous weather.

2. **Explain** Why are hurricanes hazardous?

3. **Infer** Look at the large photo. What could this type of weather do to houses built along a beach?

189

They could brack and flowd

They can damage a lot of property. Horiecan r thondersom lots of rain

Reducing the Impact of Flooding

A **flood** is an overflow of water that covers land that is usually dry. Floods can happen during short, heavy rains or when rain falls for a long time. Hurricanes can cause flooding, too. Both the heavy rains and the wind pushing ocean water onto land can cause floods during hurricanes.

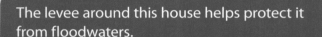
The levee around this house helps protect it from floodwaters.

The Thames River sometimes floods parts of London, England. This barrier can move large walls of steel up to hold back high water.

DCI ESS3.B: Natural Hazards. A variety of natural hazards result from natural processes. Humans cannot eliminate natural hazards but can take steps to reduce their impacts. (3-ESS3-1)
NS Science Is a Human Endeavor. Science affects everyday life. (3-ESS3-1)

By understanding how floods work, engineers can build structures to reduce their damage. A **levee** is an earthen wall. Levees slow or stop water from flooding an area. A **dam** is a concrete wall built across a river. Dams hold back water that could cause flooding. Flooding can also be prevented using bags of sand to block the water or other materials that soak up water.

This tube contains a material that expands to form a barrier against floodwater.

High water after Hurricane Katrina flooded many homes in Louisiana.

Wrap It Up! 🗒 My Science Notebook

1. **Cause and Effect** What are some causes of flooding?

2. **Explain** How do people try to reduce the damage caused by flooding?

191

Reducing the Impact of Wind

A gentle breeze might cool you on a warm day. But strong winds in tornadoes and hurricanes can cause much damage. Yet people can take action to protect their lives and property. They can protect themselves, too, from objects flung about by strong winds.

Special glass in doors and windows resists breaking as wind and flying objects blow against it. Also, roof tiles held down with nails instead of staples are not as easily blown off by the wind.

Tornadoes have stronger winds than any other kind of storm.

DCI ESS3.B: Natural Hazards. A variety of natural hazards result from natural processes. Humans cannot eliminate natural hazards but can take steps to reduce their impacts. (3-ESS3-1)
NS Science Is a Human Endeavor. Science affects everyday life. (3-ESS3-1)

People can protect themselves from strong winds by closing all of the doors in their homes. Staying in the basement or a small room without windows until the storm is over can help them stay safe, too.

Roofs that slope on all sides stand up better in high winds than roofs that slope on only two sides.

Storm shelters can keep people safe during a tornado.

Wrap It Up!

1. **Identify** Name two storms with strong winds.

2. **Explain** Tell how people can protect property from strong winds.

3. **Describe** How can people protect themselves during a storm with strong winds?

TEM SCIENCE
TECHNOLOGY
ENGINEERING
MATH

EERING PROJECT

sign a
d-Resistant Tower

e top of a tall tower you can see far. On a
ay, the view is amazing! But what happens in
Powerful winds can make a tower sway. If
er is not strong enough, it might even break.
s must consider the effects of severe weather
ey design towers and other structures.

wants to build a new tower. The tower must
hough for people to see the whole city. It
o be able to stand up against the force of the
u will help the city. You will make a model of
that can resist strong winds.

3–5-ETS1-3. Plan and carry out fair tests in which variables are controlled and failure points are
nsidered to identify aspects of a model or prototype that can be improved.

The Challenge

Your challenge is to design and build a model of a tower that can stand up in a strong wind. Your tower must:

- be at least 30 centimeters tall
- hold up a tennis ball
- withstand wind from a fan for 30 seconds

The Space Needle in Seattle, Washington, is designed to withstand winds up to 322 kilometers per hour (200 miles per hour).

STEM
ENGINEERING PROJECT
(continued)

The Space Needle is designed so that it can sway slightly in the wind.

1 Define the problem. My Science Notebook

Think about the problem you are solving. What does your tower need to do? Those things are the criteria of the problem. Criteria tell if your design is successful.

Your teacher will give you materials for building your tower. You cannot use any other materials. Your tower must be able to stand on its own. You cannot tape it in place.

Look at the fan you will be using to test your tower. Turn it on to get a sense of the force of its wind.

Write the problem you need to solve in your notebook. List the criteria and the constraints.

2 Find a solution.

Look at the materials. How can you use them to make a wind-proof tower? Think about these questions:

- What materials can help make your tower tall?
- What shape will be best at resisting the wind? Round? Square? Triangular?
- How will you keep your tower from tipping over?

Be creative! Sketch your design. Share it with your team. Discuss your teammates' designs. Choose the best design. Use other ideas to make it better. Draw your final design.

The wide base of the Eiffel Tower in Paris, France, gives it strength.

3 Test your solution.

Build your tower. Can it hold a tennis ball? Adjust your tower to make it as strong as you can. Make notes to show what you change. Adjust your drawing.

Use a ruler to measure your tower. How tall is it?

Place your tower 30 centimeters from the fan. Blow the fan on it for 30 seconds. Write your observations in your notebook.

Discuss the wind test with your team. Talk about the other criteria. Did your design solve the problem?

4 Refine or change your solution.

Discuss ways to improve your tower. Use your ideas to change it.

Test your tower the same way you did in Step 3. Place your tower 30 centimeters from the fan. Make the test fair. Did your changes improve your tower?

Present your tower to the class. Tell the results of your tests. Answer questions about your design. Ask questions about the other teams' designs.

How could you improve your design? Record your ideas in your notebook.

Reducing the Impact of Lightning

Lightning is a bright electrical discharge. Lightning can flash between parts of a cloud, from the ground to a cloud, or from a cloud to the ground. A lightning strike can cause fires. People need to avoid being struck, too.

The best way to protect yourself from lightning is to stay indoors during thunderstorms. Once inside, stay away from windows, water, and electronic equipment. If you must be outside during a storm, stay away from water and from tall objects in the area.

Tall buildings are built to direct the electricity from lightning safely into the ground.

DCI **ESS3.B: Natural Hazards.** A variety of natural hazards result from natural processes. Humans cannot eliminate natural hazards but can take steps to reduce their impacts. (3-ESS3-1)
NS Science Is a Human Endeavor. Science affects everyday life. (3-ESS3-1)

Many structures are protected from lightning by lightning rods. A lightning rod does not stop lightning. Instead, it safely guides the electricity from a strike along a wire to the ground.

Lightning rods can help prevent fires and other damage.

Lightning strikes the metal rod.

Electricity travels along a wire from the rod to the ground.

Electricity harmlessly enters the ground.

Wrap It Up!

My Science Notebook

1. **Describe** Why is lightning dangerous?

2. **Explain** What should you do if you are caught outdoors during a thunderstorm?

3. **Infer** Many people unplug their electronic equipment when a thunderstorm is predicted. Why do you think this is so?

199

Make a Claim

You know that hurricanes can cause a lot of damage. Engineers work to design structures that will stand up to the wind and rain of a hurricane. In this activity, you'll be the engineer. You will design a house for an area where hurricanes are common. You will present the design to others and explain why it will be successful. Then you will use feedback to change your design to make it even better!

1. **Define the problem.** My Science Notebook

   How can you design a house to protect it from hurricane damage?

2. **Find a solution.**

   Review what you know about hurricanes. What type of weather do they bring to an area? What kinds of damage do hurricanes cause? How can this damage be reduced or prevented? Do research or look back at other lessons. Decide on the solution you think would be best to protect a house from hurricane damage. What are the risks of the solution? What are its benefits?

   Draw and label your solution. Write how your solution would work.

3. **Defend your solution.**

   Present your solution to your class. State why you think your solution best meets the needs of people. Support your claim with evidence. Listen to how your classmates say you might improve your solution.

4. **Refine or change your solution.**

   Use the feedback from your classmates to improve your solution. Present your revised solution to the class.

PE 3-ESS3-1. Make a claim about the merit of a design solution that reduces the impacts of a weather-related hazard.
NS Science Is a Human Endeavor. Science affects everyday life. (3-ESS3-1)

You can think like an engineer and identify a solution that will help a house stand up to the force of a hurricane.

Severe-Storms Researcher

As a boy, Justin Walker was fascinated by tornadoes. He grew up in Oklahoma, a state that some might call "storm central." Once, a deadly tornado tore through his town. It caused terrible damage. Justin was in awe. He knew then that he wanted to study tornadoes someday and help keep people safe. Justin followed his passion and became a meteorologist.

As an adult, Justin chases storms with teams of other experts. Special equipment on their vehicles gathers data about the storms. Justin has followed hundreds of storms. The work is dangerous, cold, and wet. But it is also rewarding. Justin enjoys working on the equipment as well as studying the tornadoes. These teams' brave research helps people better prepare for severe storms.

Justin Walker worked, at times, with National Geographic Explorer Tim Samaras. Tim is a legend in storm research. In his time, Tim was a passionate storm chaser and engineer. He designed new instruments to better study tornadoes. In 2013, Tim was killed by a tornado. Tim's important work lives on and helps keep others safe.

DCI ESS3.B: Natural Hazards. A variety of natural hazards result from natural processes. Humans cannot eliminate natural hazards but can take steps to reduce their impacts. (3-ESS3-1)
NS Science Is a Human Endeavor. Science affects everyday life. (3-ESS3-1)

Justin Walker is a severe-storm meteorologist. His work chasing storms is dangerous but important. He has been featured on television programs and films, including the National Geographic IMAX film, *Extreme Weather*.

The equipment mounted on this truck helps scientists better understand tornadoes.

Check In

My Science Notebook

Congratulations! You have completed *Earth Science.* Let's reflect on what you have learned. Here is a checklist to help you judge your progress. Look through your science notebook to find examples of each item in the list.

What could you do better? Write it on a separate page in your science notebook.

▼ Read each item in this list. Ask yourself if you think you did a good job of it.

For each item, select the choice that is true for you: A. Yes B. Not Yet

- I defined and drew pictures of science vocabulary, science concepts, and main ideas.
- I labeled drawings. I included captions and notes to explain ideas.
- I collected objects, such as photos and magazine or newspaper clippings.
- I used tables, charts, or graphs to record observations and data in investigations.
- I recorded evidence for explanations and conclusions in investigations.
- I described how scientists and engineers answer questions and solve problems.
- I asked new questions.
- I did something else. (Tell about it.)

Reflect on Your Learning My Science Notebook

1. In what ways were you creative during the investigations you conducted?

2. Choose one science idea that you think was most important to learn about. Explain your thinking.

More to Explore

As a boy, I was inspired by the stories I heard of the Boiling River. I began to ask questions about something I didn't understand. When I met the local people, I learned that they, too, had wondered what could make the river so hot. It showed me that people have always looked for ways to explain the world around us. The records I kept in my notebook helped me to do that.

Write about how you used your notebook. Tell how it helped you ask questions and find answers about the natural world. What were some of the most interesting and surprising things you learned? What were some consistent patterns you noticed? Discuss what you learned with classmates. And remember, as you continue *Exploring Science,* never be afraid to ask new questions. Always be curious!

Andrés Ruzo can't always visit the Boiling River, but he can look back at his notebooks as he continues to ask questions and explore the natural world.

Science Safety

Be responsible, look, and listen in the science lab. Follow all lab safety rules to stay safe. Know what procedure to follow for each lab you conduct. If anything is unclear, ask an adult for help. Always be aware of your space. Report an accident to an adult immediately. Tell your teacher about any allergies you have.

The Lab Space

- Know where the first aid kit is kept.
- Know the location of the fire blanket.
- Sit only on lab chairs or stools, never lab tables.
- Do not run.
- Keep your area neat.

Lab Clothing

- Tie back loose or long hair.
- Do not let jewelry or clothing hang loosely.
- Shoes should cover the foot; no sandals.
- Wear goggles, gloves, or a lab apron when told.

Animal and Plant Safety

- Be aware of the living things in your lab.
- Wash hands after handling plant or animal material.

Chemical Safety

- Do not eat or drink anything in the science room.
- Never mix chemicals.
- Find out where to dispose of chemicals.
- Keep hands away from your eyes and mouth.
- Wash your hands when finished.

Fire and Electrical Safety

- Do not touch charged ends of batteries.
- Never connect multiple batteries in a circuit.

Glass Safety

- Ask an adult to place broken glass in a sealed container.

Cleanup

- Close all containers.
- Return materials to their correct storage locations.
- Throw out used gloves.
- Wash hands with soap and water.

Tables and Graphs

When scientists and engineers make observations, they often record different kinds of information. This is called data. Data can be measurements or facts that tell about observations.

Data help scientists and engineers answer questions, make predictions, and ask new questions. Scientists and engineers need to organize their data to make sense of the data and share with others. Tables and graphs are tools for organizing, summarizing, and sharing data. You can use these tools to organize data, too.

Tables

A table is a set of **rows** and **columns.** Rows and columns set up a simple grid that makes a place for every piece of information.

Column

Row

PRECIPITATION IN WESTERN U.S. CITIES	
City, State	**Precipitation (cm (in.))**
Las Vegas, Nevada	10 cm (4 in.)
Phoenix, Arizona	20 cm (8 in.)
Riverside, California	25 cm (10 in.)
San Diego, California	25 cm (10 in.)
Los Angeles, California	30 cm (12 in.)
Denver, Colorado	38 cm (15 in.)
San Jose, California	40 cm (16 in.)
Salt Lake City, Utah	40 cm (16 in.)

Source: National Weather Service, 2016

This table has two columns. One column shows cities and states. The other column gives amounts of precipitation.

Cite the source of your data, if you did not collect it yourself. This shows others how to confirm whether your data are valid, or reliable.

A table is a useful way to organize a list of information. You can add new information to a table by adding more rows or columns. What information does this table help organize? How is it organized?

Bar Graphs

A bar graph is useful for comparing different amounts of similar information.

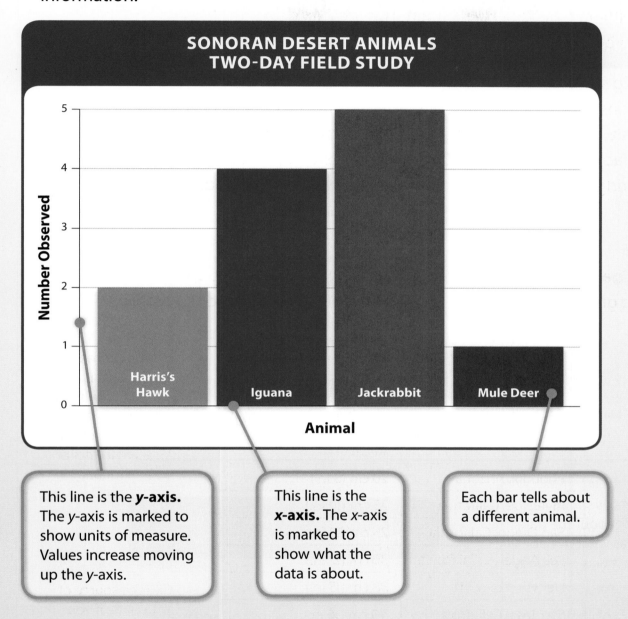

**SONORAN DESERT ANIMALS
TWO-DAY FIELD STUDY**

Number Observed

Harris's Hawk

Iguana

Jackrabbit

Mule Deer

Animal

This line is the **y-axis.** The y-axis is marked to show units of measure. Values increase moving up the y-axis.

This line is the **x-axis.** The x-axis is marked to show what the data is about.

Each bar tells about a different animal.

Each bar is a different height. Bar height answers the question "How many?" The top of each bar lines up with a unit of measure on the y-axis. Which animal species was observed the most on a visit to the Sonoran Desert?

A pictograph uses a series of pictures instead of bars. What pictures might you use to make a pictograph of this information?

Line Graphs

A line graph is good for showing how data changes over time.
Like a bar graph, a line graph has an *x*-axis and a *y*-axis.

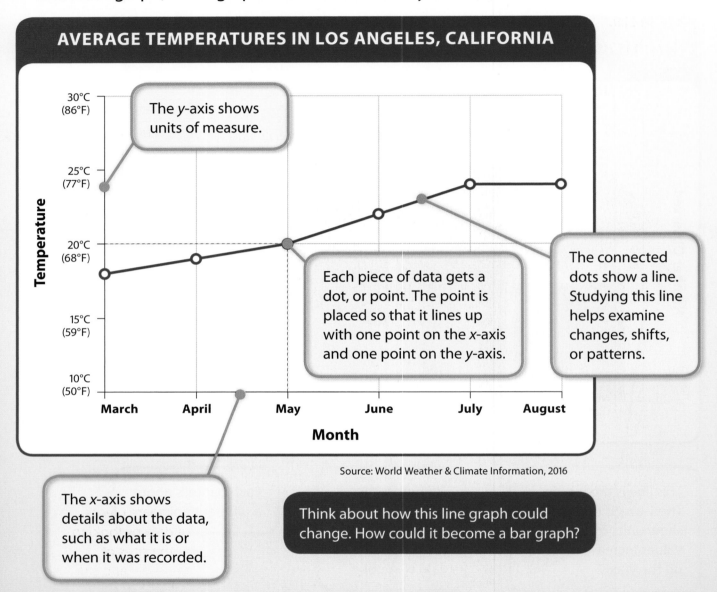

AVERAGE TEMPERATURES IN LOS ANGELES, CALIFORNIA

The *y*-axis shows units of measure.

Each piece of data gets a dot, or point. The point is placed so that it lines up with one point on the *x*-axis and one point on the *y*-axis.

The connected dots show a line. Studying this line helps examine changes, shifts, or patterns.

The *x*-axis shows details about the data, such as what it is or when it was recorded.

Source: World Weather & Climate Information, 2016

Think about how this line graph could change. How could it become a bar graph?

A line graph might look at data across time, across ages, or across distances. This line graph shows how temperature changes across several months. In what month does the coolest temperature occur? What pattern do you see across the months?

Circle Graphs

Circle graphs show data as parts that make up a whole. These graphs are also called "pie graphs" because the parts look like pie slices. The whole circle stands for the total amount, or 100%, of the data collected. A circle graph is useful for data that tell how much or how many of something.

ENERGY RESOURCES IN THE UNITED STATES

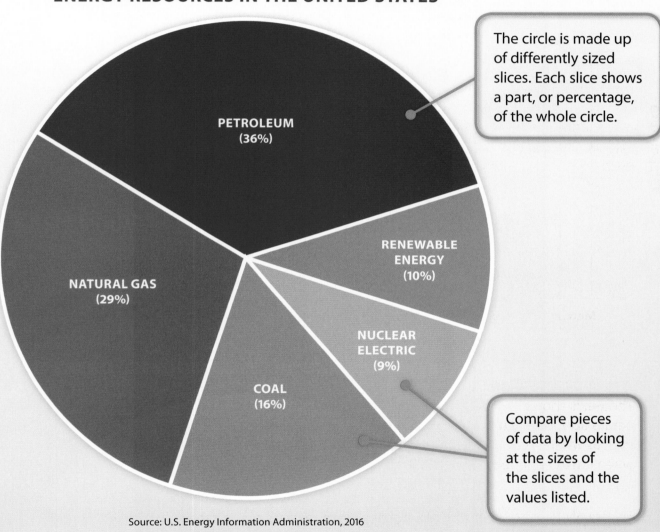

The circle is made up of differently sized slices. Each slice shows a part, or percentage, of the whole circle.

PETROLEUM (36%)

RENEWABLE ENERGY (10%)

NATURAL GAS (29%)

NUCLEAR ELECTRIC (9%)

COAL (16%)

Compare pieces of data by looking at the sizes of the slices and the values listed.

Source: U.S. Energy Information Administration, 2016

The blue slices in this graph show how much of the total use came from nonrenewable energy resources. The green slice shows how much came from renewable energy resources. Which energy resource is used the most? Which energy resource is used the least? How much of the energy used came from nonrenewable resources?

Glossary

A

acquired trait (uh-KWĪ-urd TRĀT)
An acquired trait is a trait an organism gains from the environment. (p. 88)

air pressure (AIR PRE-shur)
Air pressure is the force with which air pushes on Earth. (p. 171)

anemometer (an-uh-MOM-i-tur)
An anemometer is a tool that measures wind speed. (p. 172)

attract (uh-TRAKT)
Objects that attract each other exert a pull without having to touch. (p. 44)

B

balanced forces (BA-lenst FOR-sez)
Balanced forces cancel each other out and do not cause an object to move. (p. 24)

barometer (bah-RAHM-i-tur)
A barometer is a tool that measures air pressure. (p. 170)

biodiversity (bio-di-VUR-site)
Biodiversity is the variety of living things on Earth. (p. 109)

C

climate (KLĪ-mit)
Climate is the general pattern of weather in an area over a long period of time. (p. 182)

column (KO-lum)
A column is a vertical section of a table. (p. 208)

The **climate** in Coca, Ecuador is hot and humid all year.

213

D

dam (DAM)
A dam is a wall built across a river to hold water back. (p. 191)

deciduous (dē-SID-yū-us)
Deciduous trees shed their leaves in the fall. (p. 115)

E

ecosystem (Ē-kō-sis-tum)
An ecosystem is all the living and nonliving things in an area and the ways they interact. (p. 106)

evidence (EV-i-dens)
A piece of evidence is an observation that supports an idea or conclusion. (p. 10)

experiment (eks-PAIR-i-ment)
In an experiment, you change only one variable, measure or observe another variable, and control other variables so they stay the same. (p. 13)

F

flood (FLUD)
A flood is an overflow of water that covers land that is usually dry. (p. 190)

force (FORS)
A force is a push or a pull. (p. 22)

fossil (FAH-sil)
A fossil is a preserved trace of an organism that lived long ago. (p. 141)

front (FRUNT)
A front is a place where two very large masses of air meet. (p. 174)

H

habitat (HA-bi-tat)
A habitat is the place where a plant or animal lives and gets everything it needs to survive. (p. 116)

hibernate (HĪ-bur-nāt)
While an animal hibernates in winter, its body does not use much energy. (p. 115)

I

infer (in-FUR)
When you infer, you use what you know and what you observe to draw a conclusion. (p. 10)

inherited trait (in-HAIR-it-ed TRĀT)
Traits that are passed down from parents to offspring are inherited traits. (p. 82)

investigate (in-VES-ti-gāt)
You investigate when you carry out a plan to answer a question. (p. 12)

L

larva (LAHR-vuh)
A larva is a young, wingless form that hatches from the egg of many insects. (p. 75)

levee (LE-vē)
A levee is an earthen wall that holds back water. (p. 191)

life cycle (LĪF SĪ-kul)
A life cycle is a series of changes a living thing goes through during its lifetime. (p. 70)

Inherited traits cause offspring to look like their parents.

M

magnet (MAG-net)
A magnet is a material that is able to push or pull certain kinds of metal without contact. (p. 44)

magnetic force (mag-NE-tik FORS)
Magnetic force is the pull or push that magnets exert. (p. 44)

migrate (MĪ-grāt)
To migrate means to move to a different place to meet basic needs. (p. 112)

model (MO-del)
In science, models are used to explain or predict phenomena. A model can show how a process works in real life. (p. 13)

N

net force (NET FORS)
All the forces acting on an object add up to the net force. (p. 24)

O

observe (ub-ZURV)
When you observe, you use your senses to gather information about an object or event. (p. 12)

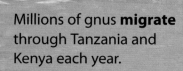

Millions of gnus **migrate** through Tanzania and Kenya each year.

P

pack (PAK)
A pack is a group of closely related animals that live and hunt together. (p. 132)

pole (PŌL)
Poles are the places on a magnet where its pull or push is the strongest. (p. 44)

precipitation (prē-sip-uh-TĀ-shun)
Precipitation is water, such as rain or snow, that falls from the clouds. (p. 171)

predator (PRE-di-tur)
A predator is an animal that hunts other animals for food. (p. 117)

prey (PRĀ)
An animal that is hunted as food by other animals is prey. (p. 133)

pupa (PYŪ-puh)
A pupa is the life cycle stage an insect goes through between being a larva and becoming a mature adult. (p. 75)

R

rain gauge (RĀN GĀJ)
A rain gauge is a tool that measures the amount of rain that falls. (p. 170)

regular motion (REG-yū-lur MŌ-shun)
Regular motion is motion that repeats in a pattern. (p. 36)

repel (rē-PEL)
Objects that repel each other exert a push without touching. (p. 44)

reproduce (rē-prō-DŪS)
When organisms reproduce, they make more of their own kind. (p. 70)

row (RO)
A row is a horizontal section of a table. (p. 208)

S

school (SKŪL)
A school is a group of many fish that swim close together. (p. 134)

static electricity (STA-tik i-lek-TRI-suh-tē)
Static electricity is an electric charge that builds up in a material. (p. 54)

swarm (SWORM)
A swarm is a large group of insects moving together from one place to another. (p. 136)

T

tadpole (TAD-pōl)
A tadpole is the young stage of a frog or toad that lives in water and has not yet developed legs or lungs. (p. 77)

thermometer (thur-MAH-mi-tur)
A thermometer is a tool that measures temperature. (p. 170)

trait (TRĀT)
A trait is a characteristic of a living thing. (p. 82)

U

unbalanced forces (un-BA-lenst FOR-sez)
Unbalanced forces on an object cause the object to move. (p. 27)

V

variable (VAIR-e-u-bl)
A variable is a factor that can change or be controlled in an experiment, investigation, or model. (p. 13)

variation (vair-ē-Ā-shun)
Variations are differences in the traits of the same types of living things. (p. 101)

Fish swim in a **school** for protection.

W

weather (WE-thur)
Weather is what conditions in the air are like outside at a given time and place. (p. 168)

wind (WIND)
Moving air is called wind. (p. 171)

wind vane (WIND VĀN)
A wind vane is a tool that shows the direction from which the wind is blowing. (p. 170)

X

x-axis (EKS-AKSIS)
The x-axis on a graph is the horizontal base line. The x-axis is usually presented on the bottom of the graph. (p. 209)

Y

y-axis (WI-AKSIS)
The y-axis on a graph is the vertical base line. The y-axis is usually presented on the left side of the graph. (p. 209)

The **wind** is strong enough to keep this leaning person from falling.

Index

Giza, Egypt, 187
Golden Gate Bridge, 28–31
Grand Teton National Park, 176
Graphic organizers, 96
Graphs, using, 4–5, 67, 109, 178–181
Grasses, 111, 112–113
Grasslands, 112–113
Gravity, 24
Great frigatebirds, 102–103
Green River Formation, 144, 150

H

Habitats
    beaver pond, 116–117
    evidence of in fossils, 144–145
    human activities changing, 118–125
    needs of living things provided by,
    152–159
    rooftop gardens, 125
    seasonal changes, 112–113
    in space, 120
Hand lens, 108
Hatfield, Samantha Chisholm,
    184–185
Hawks, 134, 156
Herds, 130, 134
Hibernation, 115
Honey, 115
Horner, Jack, 142–143
Hot springs, 3
Houses, 118, 124
Human activities
    changing ecosystems, 124–125
    changing land, 118–119
Hummingbirds, 156
Hump of a camel, 155
Hunting, 132–133
Hurricane Katrina, 190–191
Hurricane Sandy, 188–189
Hurricanes
    damage from flooding, 188–191
    damage from winds of, 192–193
    reducing the impact of, 192–193,
    200–201
Hyenas, 133

I

Inferences, 10, 16–17
Inherited behavior, 94

Inherited traits
    functions, 86–87
    looks, 82–83, 104–105
Insects
    army ants, 133
    eel ants, 185
    effects of forest fires on, 111
    entomologists' study of, 84–85
    habitats of, 125
    life cycle of ladybugs, 74–75
    plants provided for, 118
    thorn bugs, 101
Instinct, 94
International Space Center, 120–121
Investigate
    Electric Forces, 56–57
    Electromagnets, 48–49
    Environment and Traits, 98–99
    Fossils, 16–17, 148–149
    Life Cycles, 78–79
    Magnetic Force, 46–47
    Motion, 38–39
    Weather, 172–173

J

Jalapeño pepper plant, 72–73
John Day Dam, 127
Jurassic Park, 142

K

Katydids, 104–105
Key on maps, 174–175
Kinshasa, Democratic Republic of
    Congo, 187
Kovalevskaya, Sofia, 40–41

L

Ladybugs, 74–75
Land, 118–125
Larva, 75
Laws of nature, 7
Learned behaviors, 92–93
Leaves, 115
Leopard frogs
    habitat of, 154–155
    life cycle of, 76–77
Let's Explore! (Notebook)
    observations, 164–165
    patterns of motion, 18–19
    scientific models, 66–67
Levee, 190–191

Life cycles
    of jalapeño pepper plants, 72–73
    of ladybugs, 74–75
    of leopard frogs, 76–77
    of orangutans, 70–71
    sequencing the stages of, 80–81
    of spotted salamander, 78–79
Life Science, 66–163
Light, 90, 156–157
Lightning
    reducing the impact of, 198–199
    during thunderstorms, 188–189
Lightning rods, 199
Lions, 94, 133, 134
Living things
    biodiversity of, 109
    changes to environments, 116–117
    changes with seasons, 114–115
    effects of forest fires on, 110–111
    in forests, 106–107
    human activities changing
    environment of, 118–119
    inherited traits of, 82–87
    life cycles of, 70–81
    meeting needs in environments,
    152–159
    variations and survival, 100–101
    See also Animals; Plants
Louisiana, 190–191
Lungs, 154

M

Mackerel, 134–135
Magadan, Russia, 187
Magnetic force
    attracting and repelling objects,
    44–47
    electromagnets, 48–49
    investigation of, 58–59
Magnets
    attracting and repelling objects,
    44–47
    design a use for, 60–61
    electromagnets, 48–49
    investigation of, 58–59
    poles of, 44
Mallard ducks, 158
Maps
    of climates, 182–187
    of Columbia River, 126
    of environments, 150
    weather maps, 174–175
Marine ecologist, 160–161
Marine protected areas, 160

S

T

W

Walker, Justin, 202–203

Warm front, 174

Water
    animals' need of, 154–155
    effect on plant growth, 98–99
    in grasslands of East Africa, 112–113
    as nonliving thing, 106–107

Weather
    changes during the day, 169
    changes in temperature, 114–115
    changes with seasons, 136, 169, 176–179
    coping with changes in, 131
    effect on plants' traits, 90–91
    in grasslands of East Africa, 112–113
    hurricanes, 188–191
    lightning, 188–189, 198–199
    measurement of, 170–173
    organizing data, 180–181
    patterns and predictions, 174–175
    rainy and dry seasons, 112–113
    storms, 166–169
    thunderstorms, 169, 188–189

Weather maps, 174–175

Whales, 130

Wheatgrass, 98–99

Wheelchair, 22

White-bearded gnu, 112–113

Whooping cranes, 86–87

Wildebeests, 112–113, 134

Wildflowers, 111

Wind
    effect on trees, 90–91
    effect on weather, 171
    measurement of speed and direction, 170–173
    reducing the impact of, 192–193

Wind vane, 170

Wind-resistant towers, 194–197

Winter
    effect on plants and animals, 114–115
    weather in, 176–179

Wolves, 130, 132–133

Women
    climate scientist, 185
    entomologist, 84–85
    equal rights for, 40–41
    first European professor, 40–41

Writing Project, Animal Behavior, 94–97

Wyoming, 144–145, 176

Y

Young
    katydids, 104–105
    leopard frogs, 77, 154
    orangutans, 70–71
    plants, 73
    salamanders, 78

Yuma, Arizona, 182

Z

Zebras, 131, 134

126-127 (tc) ©Frans Lanting/National Geographic Creative. (bc) ©Mark Conlin/Oxford Scientific/Getty Images. 127 (tr) ©Michael Sedam/BlueMoon Stock/Alamy Stock Photo. 128-129 (spread) ©Mark Conlin/Oxford Scientific/Getty Images. 129 (cl) ©Gary Braasch/Encyclopedia/Corbis. (cr) ©Natalie Fobes/Science Faction/SuperStock. (bl) ©One Fish Engineering, LLC. (br) ©Harald Sund/Photographer's Choice/Getty Images. 130-131 (spread) ©Tony Camacho/Science Source. 131 (tr) ©Konrad Wothe/imagebroker/Alamy Stock Photo. (cr) ©Frans Lanting/National Geographic Creative. 132 ©Anup Shah/Nature Picture Library. 132-133 ©Dan Stahler/National Geographic Creative. 134-135 ©Christopher Swann/Science Source. 136-137 ©Jessica Solomatenko/Flickr Open/Getty Images. 138 (tr) ©Mapping Specialists. 138-139 (spread) ©Georgesanker/Alamy Stock Photo. 139 (bl) ©Joel Sartore/National Geographic Creative. (br) ©Tom Uhlman/Alamy Stock Photo. 140 ©Tom Bean/Alamy Stock Photo. 140-141 ©Stocktrek Images, Inc./Alamy Stock Photo. 142 (tl) ©National Geographic Learning. (tr) ©James L Amos/Getty Images. (b) ©Universal/courtesy Everett/Everett Collection. 143 ©National Geographic Learning. 144 ©Mapping Specialists. 144-145 ©John Cancalosi/Alamy Stock Photo. 145 ©Brad Mitchell/Alamy Stock Photo. 146 (tc) ©Mapping Specialists. 146-147 (spread) ©Kevin Schafer/Alamy Stock Photo. 147 (cr) ©Steve Austin/Art Directors/Art Directors & TRIP/Alamy Stock Photo. 147 (cl) ©National Geographic Learning. (cr) ©National Geographic Learning. (bl) ©National Geographic Learning. (br) ©National Geographic Learning. (bc) ©National Geographic Learning. 148-149 (spread) ©Greg Dale/National Geographic Creative. 149 (tl) ©National Geographic Learning. (tr) ©National Geographic Learning. 150 (tr) (bc) ©John Cancalosi/National Geographic Creative. 150-151 (spread) ©Mapping Specialists. 151 (tl) ©O. Louis Mazzatenta/National Geographic Creative. (tr) ©Darlyna A. Murawski/National Geographic Creative. (cr) ©Frans Lanting/National Geographic Creative. (bl) ©John Cancalosi/National Geographic Creative. 152-153 (spread) ©Robert Postma/Design Pics/Getty Images. 153 (tr) ©Craig K. Lorenz/Science Source. 154-155 (spread) ©Michael Durham/Minden Pictures. 155 (cr) ©Frank Lukasseck/Photographer's Choice/Getty Images. 156-157 (spread) ©Dante Fenolio/Science Source. 157 (tr) ©Darrell Gulin/Danita Delimont/Alamy Stock Photo. (cr) ©Dante Fenolio/Science Source. 158 (br) ©Gregory/Photri Images/Alamy Stock Photo. 158-159 (spread) ©Ken Catania/Visuals Unlimited/Corbis. 159 (b) ©Ingo Arndt/Minden Pictures. 160 (tr) ©Enric Sala/National Geographic Creative. 160-161 (spread) ©Josep Clotas. 161 (tr) ©Rebecca Hale/National Geographic Creative. 163 ©Andrés Ruzo/National Geographic. 164 (tr) ©Sofía Ruzo. 164-165 ©Nuno Filipe Pereira/EyeEm/Getty Images.

Earth Science: Weather and Climate

166-167 ©Mark Duffy/National Geographic Creative. 168 (cl) ©Morey Milbradt/Stockbyte/Getty Images. (bl) ©Steele Burrow/Alamy Stock Photo. 168-169 (spread) ©David Parsons/E+/Getty Images. 170 (cl) ©Jaywarren79/Shutterstock.com. (cl) ©txking/Shutterstock.com. (cl) ©Victor Torres/Shutterstock.com. (bl) ©BMJ/Shutterstock.com. 170-171 ©epa european pressphoto agency b. v./Alamy Stock Photo. 172 (bl) (cl1) ©National Geographic Learning. (cl2) ©National Geographic School Publishing. (bl) (bc) (cr) (br) ©National Geographic Learning. (cr) ©National Geographic Learning. (br) ©National Geographic Learning. 172-173 (spread) ©Christopher J. Bandera/Flickr Open/Getty Images. 173 (tl) ©National Geographic Learning. (tr) ©National Geographic Learning. 174-175 (spread) ©Inigo Cia/

Flickr Open/Getty Images. 175 (t) ©Mapping Specialists. (b) ©Mapping Specialists. 176 (tr) ©Mapping Specialists. (bl) ©Lorenz Britt/Alamy Stock Photo. (br) ©Dennis Flaherty/Science Source. 176-177 (spread) ©John Lamb/The Image Bank/Getty Images. 177 (tr) ©Enrique R Aguirre Aves/Getty Images. (tl) ©Russell Burden/Photolibrary/Getty Images. 178 (t) ©Mapping Specialists. 178-179 (spread) ©Dick Durrance II/National Geographic Creative. 180-181 ©NoDerog/E+/Getty Images. 182 (bl) ©Anna Gorin/Flickr/Getty Images. (bc) ©Richard Cummins/Robert Harding World Imagery/Corbis. 182-183 ©Mapping Specialists. 183 (cr) ©James Schwabel/Alamy Stock Photo. (tl) ©AP Images/The Post-Standard/Nicholas Lisi. 184 (t) ©Cyril Ruoso/Nature Picture Library. (b) ©Minden Pictures. 185 ©National Geographic Learning. 186 (bl) ©Geoffrey Kuchera/Shutterstock.com. (tr) ©Colin Palmer Photography/Alamy Stock Photo. (bc) ©Pawel Toczynski/Photolibrary/Getty Images. 186-187 (spread) ©Mapping Specialists. 187 (tr) ©antonio_mag/Shutterstock.com. (tc) ©Taylor Kennedy-Sitka Productions/National Geographic Creative. (cr) ©John White/Flickr Open/Getty Images. (bl) ©Skip Brown/National Geographic Creative. 188 (tr) ©Dale O'Dell/Alamy Stock Photo. 188-189 (spread) ©AP Images/Gerry Broome. 189 ©James Nesterwitz/Alamy Stock Photo. 190 (cl) ©Scott Olson/Staff/Getty Images News/Getty Images. (bl) ©Amelia Ayaoge/Alamy Stock Photo. 190-191 (spread) ©Scott Saltzman/Bloomberg/Getty Images. 191 (tr) ©Absorbent Specialty Products, LLC. 192 (t) ©National Geographic Learning. 192-193 (spread) ©Jim Reed/Corbis. 193 (t) ©National Geographic Learning. (tl) ©Lisa LeValley/Allstar Roof Systems, Inc.(tr) ©Joe Raedle/Getty Images. 194-195 ©FogStock LLC/FogStock LLC/Superstock. 196 (t) ©Radius/Radius/Superstock. 197 ©Jonathan Howison/EyeEm/Getty Images. 198-199 ©Lyle Leduc/Photographer's Choice RF/Getty Images. 199 (tr) ©Precision Graphics. 200-201 ©AP Images/Mike Groll. 202 ©Mark Thiessen/National Geographic Creative. 202-203 (spread) ©Paul Samaras. 203 (tr) ©Ryan McGinnis/Alamy Stock Photo. 205 (b) ©National Geographic Learning. 206-207 ©Steve Debenport/Getty Images.

End Matter

212-213 ©Matthew Williams-Ellis/Getty Images. 214-215 ©Mark Carwardine/Photolibrary/Getty Images. 216-217 (spread) ©Winfried Wisniewski/Minden Pictures/Getty Images. 218-219 ©Daniela Dirscherl/WaterFrame/Getty Images. 220 ©Jay Dickman/National Geographic Creative.

Illustration credits

Unless otherwise indicated, all illustrations were created by Lachina, and all maps were created by Mapping Specialists.

Program Consultants

Randy L. Bell, Ph.D.
Associate Dean and Professor of
Science Education, College of
Education, Oregon State University

Malcolm B. Butler, Ph.D.
Professor of Science Education
and Associate Director; School of
Teaching, Learning and Leadership;
University of Central Florida

Kathy Cabe Trundle, Ph.D.
Department Head and Professor,
STEM Education, North Carolina State
University

Judith S. Lederman, Ph.D.
Associate Professor and Director of
Teacher Education, Illinois Institute of
Technology

**Center for the Advancement of
Science in Space, Inc.**
Melbourne, Florida

Acknowledgments

Grateful acknowledgment is given to the authors, artists,
photographers, museums, publishers, and agents for permission to
reprint copyrighted material. Every effort has been made to secure
the appropriate permission. If any omissions have been made or if
corrections are required, please contact the Publisher.

 is a registered trademark of Achieve. Neither Achieve
nor the lead states and partners that developed the
Next Generation Science Standards was involved in the production
of, and does not endorse, this product.

Photographic and Illustrator Credits

Front cover wrap ©Christian Kober/JWL/Aurora Photos. ©Scott Leslie/
Minden Pictures.
Back cover ©Sofía Ruzo.

Acknowledgments and credits continue on page 228.

For product information and technology assistance, contact us at
Customer & Sales Support, 888-915-3276
For permission to use material from this text or product, submit
all requests online at **www.cengage.com/permissions**
Further permissions questions can be emailed to
permissionrequest@cengage.com

National Geographic Learning | Cengage
1 N. State Street, Suite 900
Chicago, IL 60602

National Geographic Learning, a Cengage company, is a provider of
quality core and supplemental educational materials for the PreK-12,
adult education, and ELT markets. Cengage is a leading provider of
customized learning solutions with employees residing in nearly 40
different countries and sales in more than 125 countries around the
world. Find your local representative at **NGL.Cengage.com/RepFinder.**

Visit National Geographic Learning online at **NGL.Cengage.com/school**

ISBN: 978-13379-11665

Printed in the United States of America
Print Number: 01
Print Year: 2018